The Medieval Town Wall of Great Yarmouth, Norfolk

A geological perlustration

John F. Potter

BAR British Series 461
2008

Published in 2016 by
BAR Publishing, Oxford

BAR British Series 461

The Medieval Town Wall of Great Yarmouth, Norfolk

ISBN 978 1 4073 0286 7

© J F Potter and the Publisher 2008

The author's moral rights under the 1988 UK Copyright,
Designs and Patents Act are hereby expressly asserted.

All rights reserved. No part of this work may be copied, reproduced, stored,
sold, distributed, scanned, saved in any form of digital format or transmitted
in any form digitally, without the written permission of the Publisher.

BAR Publishing is the trading name of British Archaeological Reports (Oxford) Ltd.
British Archaeological Reports was first incorporated in 1974 to publish the BAR
Series, International and British. In 1992 Hadrian Books Ltd became part of the BAR
group. This volume was originally published by Archaeopress in conjunction with
British Archaeological Reports (Oxford) Ltd / Hadrian Books Ltd, the Series principal
publisher, in 2008. This present volume is published by BAR Publishing, 2016.

Printed in England

BAR titles are available from:

        BAR Publishing
        122 Banbury Rd, Oxford, OX2 7BP, UK
EMAIL   info@barpublishing.com
PHONE +44 (0)1865 310431
  FAX   +44 (0)1865 316916
        www.barpublishing.com

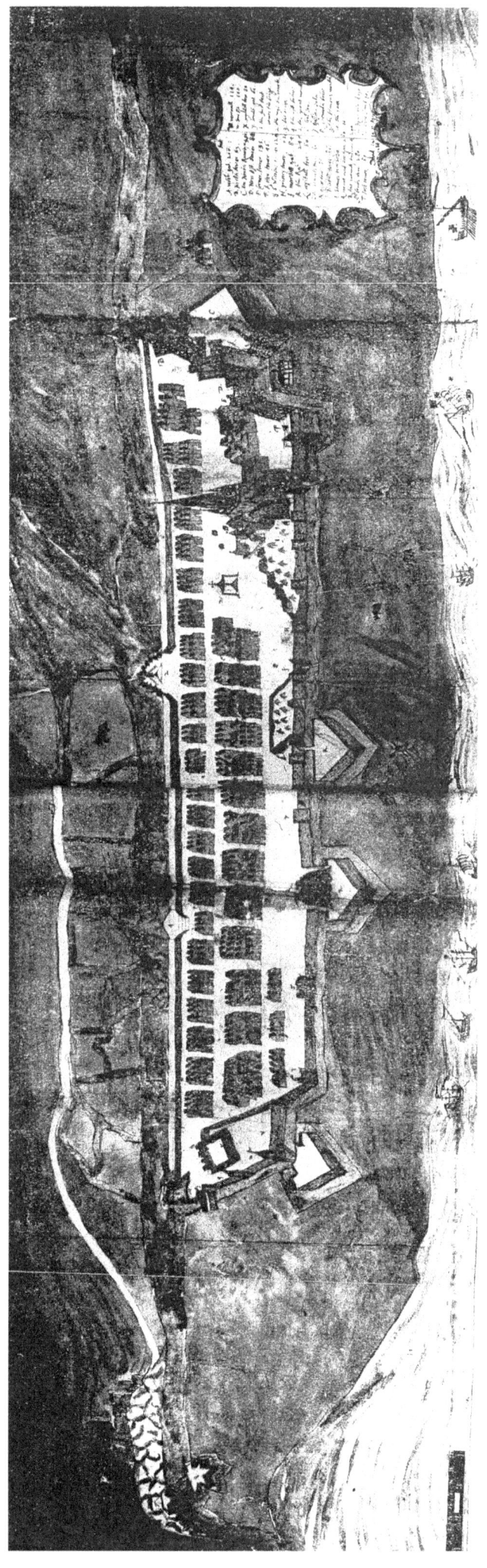

Frontispiece A plan of the Great Yarmouth fortifications as they are supposed to have appeared in 1588 and compiled at about that time. North is to the right. The plan is normally described as the Hatfield House map and the original is held by the British Museum (O'Neil and Stephens, 1942). The plan is reproduced here by kind permission from a copy held by the Norfolk Record Office. As far as can be interpreted the key reads: A – West Gate, B – North Tower, C – The North Ravely(i)n, D – Nor(th) East Tower, E – Corner Tower, F – K. Henry's Tower, G – St Nicholas Tower, H – Prieery Tower, J – Market Gate, K – the Ra(velin), L – (H)ospitall Tower, M – Bennets Tower, N – Oxne Tower, O – Pinakell Tower, P – Shanes Tower, Q – mount and curtyns, R – the mountes bulwark, S – Bowls Tower, T – East Tower, V – New wall, W – Sow Est (Tower), X – Myddell Tow(er), Y – Sowth Gate. These names are not necessarily the same as those given to the corresponding structures by subsequent authors or within the present work. Part of this plan is also reproduced as Figure 5.6.

ABSTRACT

In the Middle Ages, Great Yarmouth was a town of considerable economic and strategic significance; in 1334, it ranked fourth in English provincial towns in its wealth. At that time, such importance necessitated an elaborate protective wall. Although commenced somewhat earlier, a sustained building programme was probably not initiated until about the time confrontation with Philip VI of France precipitated England into the period of conflicts generally referred to as the Hundred Years War. The enormity and expense of the project, however, meant that the entire wall circuit was probably not completed to full intended height until the closing years of the fourteenth century.

The known historical record had revealed that Great Yarmouth was first granted authority to construct a wall in 1261. Subsequent to that date, details of murage grants, offered to provide some financial support towards the erection of the town wall, had additionally been identified. Also available were contemporary documents for restricted periods, these revealing valuable details of receipts and disbursements related to the building materials and the construction of the walls. Based on this information the wall had typically been considered to date principally from the early to mid 14th century.

This work examines in detail the construction and, more especially, material composition of the Great Yarmouth town walls. In geological terms it comprises a study set within a geographical area where recognised quality building stone, and particularly stone suitable for ashlar work, is absent. Just two principal building materials, flints from the Chalk and bricks, had previously been identified.

The examination of the wall fabric has demonstrated that the analysis outlined above was far too simplistic. The bricks within the walls were generally reused, and the flints were variable in form and style of use. Little of the very earliest wall could be identified, and repairs and replacements prevailed.

The historical records, especially those from the town's Assembly, had indicated that in the mid 16th century, it had proved necessary to modify the town's defences. This was in consequence to the potential danger of attack from the Spanish Armada and the necessity to address the new techniques of warfare involving artillery. The inside of the original walls were embanked with earth, or rampired, to provide additional strength. This study reveals that these actions actually helped to cause the partial collapse of many portions of the wall. Some of the walls, therefore, although re-incorporating this earlier wall material, were actually reconstructed about the time of, or subsequent to, the rampiring in the mid 16th century.

These studies of the variations in the composition of different stretches of the walls, therefore, reveal a complex history of repair and modification. On the evidence of the limited number of places where the wall core was exposed, very little has escaped rebuilding. However, the bricks have been especially instructive in providing a *terminus post quem* for the date particular stretches of wall were constructed, repaired or rebuilt. It has proved possible also to identify different styles of flint facework in the walls, and relate these to broad periods of workmanship. Their identification should valuably assist in the relative dating of other medieval flint walls elsewhere in the south-east of England.

The study has, wherever existing accounts are available, tried to link these records to the information obtained from the scrutiny of the walls. This has revealed the significance that the wall, although never seriously used 'in anger', has had, especially economically, upon the inhabitants of the town. Like the town's havens, the wall, its building and frequent modifications, have always drawn heavily upon the town's resources.

As early as 1261, the requirement was for a wall and a ditch. The wall, even if frequently rebuilt, still stands; a ditch adjacent to the wall may possibly once have existed. The record provided by 17th to 19th century historians is, however, of a moat in this position, capable of transporting reasonably sized boats. Its possible presence is discussed in this work and generally negated; although the occurrence, at the onset of the Civil War, of a moat of such dimensions more distant from the wall and partially across the isthmus, is not in doubt.

CONTENTS

List of Tables iii
List of Figures iv
List of Old Maps and plans (by date order) xi

ACKNOWLEDGEMENTS xiii

PROCEDURES xiv

CHAPTER ONE. GREAT YARMOUTH: GEOGRAPHICAL, HISTORICAL AND ECONOMIC BACKGROUND
1.1 Introduction 1
1.2 The foundation and development of Great Yarmouth 1
    1.2.1 Site and situation 1
    1.2.2 Brief economic background 5
1.3 Published and manuscript histories of Great Yarmouth 7
1.4 The building of Great Yarmouth town wall 8
    1.4.1 Introduction 8
    1.4.2 The raising of funds 9
    1.4.3 The construction timetable 9
    1.4.4 Wall building completion date..9
    1.4.5 Repairs to the town wall 11
1.5 Summary 14

CHAPTER TWO. THE STRUCTURE AND COMPOSITION OF THE WALLS
2.1 Introduction 15
2.2 Possible construction of the walls – earlier proposals 15
2.3 Building materials currently evident in the walls 15
    2.3.1 Flint 17
    2.3.2 Bricks..20
    2.3.3 Reused materials 22
    2.3.4 Other rock types 22
2.4 The architecture of the wall 23

CHAPTER THREE. THE TOWN WALL: 'A MODERN GEOLOGICAL PERLUSTRATION'
3.1 Introduction 26
3.2 A description of the wall 26
    3.2.1 Stretch 'A', North West Tower and wall fragments 26
    3.2.2 Stretch 'B', Town Wall Road, North East Tower, King Henry's Tower.............................32
    3.2.3 Stretch 'C', King Henry's Tower to Fishers Court 34
    3.2.4 Stretch 'D', St Nicholas School, Hospital Tower, Market Gate 37
    3.2.5 Stretch 'E', Market Gates Shopping Area 41
    3.2.6 Stretch 'F', Regent Street, Pinnacle Tower to Shave's Tower 42
    3.2.7 Stretch 'G', Shave's Tower, New Gate and the East Mount 43
    3.2.8 Stretch 'H', south of East Mount, Little Mount Gate and Harris's Tower 44
    3.2.9 Stretch 'J', White Lion Gate to Garden Gate (Alma Road)..46
    3.2.10 Stretch 'K', Garden Gate to the South East Tower..47
    3.2.11 Stretch 'L', South East Tower to Blackfriars' Tower..50
    3.2.12 Stretch 'M', Blackfriars' Tower, Palmer's Tower and the South Mount 56

CHAPTER FOUR. THE WALL FABRIC – AN ANALYSIS
4.1 Introduction..60
4.2 The Bricks 60
4.3 Exotic rock types including reused material from earlier buildings 67
    4.3.1 Reused rock material from earlier buildings...67
    4.3.2 Exotic rock material...68
4.4 Flints..68
    4.4.1 Introduction 68
    4.4.2 Flint wall face styles 69

4.5 Summary 72

CHAPTER FIVE. OTHER DEFENSIVE SYSTEMS: EARLY WALLS, TOWERS, RAMPIRES, PSEUDO-RAVELINS AND A MOAT
5.1 Early walls..73
5.2 Towers..73
5.3 Rampires 74
5.4 Pseudo-ravelins or Mounts...78
5.5 A Moat 81
        5.5.1 An early 14th century structure 81
        5.5.2 A 16th century moat 85
        5.5.3 Summary..85

CHAPTER SIX. THE FINANCIAL PROVISION FOR WALL CONSTRUCTION
6.1 Introduction..87
6.2 Early financial provision – the wall building years 87
6.3 Wall rebuilding and modification – the 15th, 16th and 17th centuries 91
6.4 Summary 91

CHAPTER SEVEN. RECORDS OF EXPENDITURE ON WALL CONSTRUCTION, MODIFICATION AND REPAIR
7.1 Introduction..92
7.2 The purchase of bricks 92
7.3 The purchase of stones and other wall materials other than bricks 92
7.4 Payments made to persons involved in the construction or building of the walls 95

CHAPTER EIGHT. RESERVATIONS AND CONCLUSIONS
8.1 Introduction – the project 96
8.2 Answers?..96
8.3 Evidence from the wall fabric survey..97

GLOSSARY 101

REFERENCES...102

ANCILLARY BIBLIOGRAPHY 105

LIST OF TABLES

Table 1.1	A list of the dates and events which are believed to have affected the Great Yarmouth walls, their construction and alteration.	12
Table 2.1	The features of flint walling used to differentiate different stretches of the flint facing to the Great Yarmouth walls.	20
Table 2.2	The information collected in the field to indicate the relative numbers of different brick types seen in sections of the Great Yarmouth walls.	21
Table 3.1	A comparison of the wall length particulars between the years 1883-1884 and 1928 based on the first and third editions of the Ordnance Survey plans. The wall is taken as 2060m. in length.	26
Table 4.1	The distribution of the principal brick types observed primarily in the core elements of Great Yarmouth town wall.	61
Table 4.2	The distribution of the principal brick types observed in the walls of certain towers in Great Yarmouth town wall. 'Modern' bricks have been excluded from these analyses.	61
Table 4.3	The various rock types, here collectively described as 'exotics', which have been recorded almost in entirety on the external surfaces of the different stretches of the Great Yarmouth walls. The rocks fall into two main groups: those of igneous and metamorphic origin which have probably mainly arrived in the walls as ships' ballast from foreign parts and those of sedimentary origin. The sedimentary rocks make up a mixture of reused building stones (as Caen and Barnack Stones), glacially or fluvioglacially derived materials (mainly of Lower Cretaceous and Jurassic origin) and a smaller number of boulders of ballast.	65
Table 4.4	Flintface styles identified in Great Yarmouth walls, and their typical age ranges.	70
Table 5.1	A summary of the points made in the text which may be offered against the presence of a moat immediately beneath and outside the town walls.	86
Table 6.1	Great Yarmouth murage grants 1261-1462. (Compiled largely from English translations of the *Calendar of Patent Rolls*, various dates, HMSO, London: with assistance from Tingey, 1913).	88
Table 6.2	Typical receipts from murage and monies spent for Great Yarmouth (principally from Turner, 1970, Appendix B).	90
Table 6.3	Simplified details of those items and goods on which tolls or taxes could be levied under the murage grant authority held by Great Yarmouth over the period 1327 to 1338. Details mainly after Tingey (1913, 140-141).	90
Table 7.1	The purchase of bricks or *tegulae* for wall construction purposes at Great Yarmouth over the period 1336 to 1345.	93
Table 7.2	The purchase of flints and stones for wall construction purposes at Great Yarmouth over the period 1336 to 1345.	94

LIST OF FIGURES

Frontispiece A plan of the Great Yarmouth fortifications as they are supposed to have appeared in 1588 and compiled at about that time. The plan is normally described as the Hatfield House map and the original is held by the British Museum (O'Neil and Stephens, 1942). The plan is reproduced here by kind permission from a copy held by the Norfolk Record Office. As far as can be interpreted the key reads: A – West Gate, B – North Tower, C – The North Ravely(i)n, D – Nor(th) East Tower, E – Corner Tower, F – K. Henry's Tower, G – St Nicholas Tower, H – Prieery Tower, J – Market Gate, K – the Ra(velin), L – (H)ospitall Tower, M – Bennets ower, N – Oxne Tower, O – Pinakell Tower, P – Shanes Tower, Q – mount and curtyns, R – the mountes bulwark, S – Bowls, Tower, T – East Tower, V – New wall, W – Sow Est (Tower), X – Myddell Tow(er), Y – Sowth Gate. These names are not necessarily the same as those given to the corresponding structures by subsequent authors or within the present work. Part of this plan is also reproduced as Figure 5.6.

Figure 1.1	The geographical situation of Great Yarmouth. The town is situated in the National Grid square TG.	1
Figure 1.2	This early map, normally referred to as the 'Hutch Map', has been attributed to the Elizabethan period (Ecclestone and Ecclestone, 1959, 20) and is thought to depict the immediate Norfolk coastline during Anglo-Saxon times. Interpretation is made more complex by geographical north being placed at the bottom of the original map (now placed on the right). Reproduced from Ecclestone and Ecclestone (1959). The original is now held in the Norfolk Record Office (Y/C 37/1).	2
Figure 1.3	A reconstruction of the Norfolk coastline in Roman times. The approximate line of the modern coastline is indicated by a broken line and the position of Great Yarmouth by an asterisk. Some of the inland water area would have been marshlands. (Partly after Pearson, 2002).	3
Figure 1.4	The Quaternary geology in the immediate vicinity of Great Yarmouth. Holocene, Flandrian deposits (N. Denes F. = North Denes Formation) rest upon Pleistocene, Anglian deposits (Corton F. = Corton Formation).	4
Figure 1.5	A plan of Great Yarmouth in the late 13th century, copied by kind permission from Rutledge (1990).	6
Figure 1.6	The importance of the herring industry to Great Yarmouth's economy is reflected in the town's coat of arms, which has been modified over time. The present coat of arms, adopted in 1563, is illustrated in c). Reproduced from Ecclestone and Ecclestone (1959, 101).	
Figure 1.7	The walled town of Great Yarmouth and sites referred to in Chapter 1.	7
Figure 2.1	Great Yarmouth town walls and their relationships with the positions of modern localities and features referred to in this work.	16
Figure 2.2	A vertical cross-section through the Great Yarmouth town wall constructed from evidence obtained from a small excavation made in 1955 on the south side of the Market Gates shopping centre. Reproduced from Green (1970, Fig. 2).	17
Figure 2.3	A typical broken flint cobble which displays a thick, less silicified, white outer cortex. To preserve this cortex the cobble was almost certainly originally quarried or mined.	19
Figure 2.4	Broken flints typically exhibit conchoidal fractures as dispayed on the broken surface of this flint.	19
Figure 2.5	A flint cobble on which nipples are displayed on the broken surface. These are a feature of many of the broken flints in the Great Yarmouth walls. For a discussion on their origin see section 2.3.1.	19
Figure 2.6	Small conchoidal flakes of flint (created when the flint is knapped) have been inserted into the mortar between the flint cobbles in this wall face. The process in East Anglia is known as galleting, and the flakes of flint as gallets.	19
Figure 2.7	This flint wall face, which occurs in the Great Yarmouth walls in stretch 'C', displays marked horizontal breaks in the flint coursework known as building lifts. For details of their origin see section 2.3.1.	20
Figure 2.8	The detail of two Type 2 brick fragments. Straw or similar material markings, created in the clay drying process, can be observed on the face of the larger fragment. The bricks are larger than those of Type 3, illustrated in Figure 2.9.	21
Figure 2.9	Type 3 brick fragments: these are thinner bricks (one, about 40mm. thick against the scale). The brick on the right has two unusual 'nail hole' markings, possibly, but with no certainty, inserted to assess the moisture content during the drying process.	21

Figure 2.10	Arched firing bays observed to the north side of Garden Gate (stretch 'J'). In this stretch of wall the firing bays have been removed almost to the outer wall and. in the first two bays, a new brick wall infills the area which would have included the arrow slit.	24
Figure 2.11	The structure of the visible portions of Great Yarmouth town wall and key measurements, related to the wall, which have been recorded in this work.	25
Figure 3.1	The positions and the customarily used names for the towers and gates of Great Yarmouth town wall.	27
Figure 3.2	Part of Swinden's Plan of Great Yarmouth produced to accompany his work of 1772. This copy has been reproduced, with their kind permission, from The Society of Antiquaries of London (1953). The map is thought to date from 1758, rather than 1738 as proposed by O'Neil (1953).	28
Figure 3.3	Great Yarmouth town wall and the positions of wall stretches 'A' to 'M', used in the description of the walls within the present work.	29
Figure 3.4	The North West Tower in about 1818. This copy of an etching by John Sell Cotman was kindly supplied by the Norwich Castle Museum and Art Gallery.	30
Figure 3.5	The North West Tower viewed from the north. The flint facing shows considerable variation in style (see section *3.2.1*). At mid-tower height the flint courses are reasonably well squared	31
Figure 3.6	The exterior of the old North Gate and the early spire of St Nicholas Church as viewed from the north-west. The gate was demolished in 1807 and this etching appeared in Preston (1819) from which this figure is reproduced.	31
Figure 3.7	The exterior of wall stretch 'B2', viewed from Ferrier Road, displays highly modified, irregularly spaced, arrow/gun slits high in the wall. The colour changes in the flint wall facing are as much to do with pointing applied at different periods as to unlike building dates.	32
Figure 3.8	'Rat-trap' bonding seen in wall stretch 'B2' where the wall has been replaced with brickwork. This unusual, double header version almost certainly dates from about 1850.	33
Figure 3.9	The flint facework of wall stretch 'B3' just west of King Henry's Tower, where both patching and pointing add to the complexity of interpretation. This wall stretch facing Ferrier Road clearly shows marked changes in its different styles of flint dressing.	33
Figure 3.10	Detail of part of the wall face shown in Figure 3.9. At least two (possibly three) different styles are visible in this small surface area. To the left, the flints carry numerous nipples; to the bottom right, the more rounded flint cobbles have been broken differently to present a single flat to conchoidal face.	33
Figure 3.11	A view from the east of the octagonal King Henry's Tower and St Nicholas Church. From the tower the wall originally ran due south; it was removed in order to extend the graveyard towards the east.	34
Figure 3.12	These randomly placed and well spaced, broken, rounded, flint beach cobbles are set into part of the south-east wall of King Henry's Tower. Nipples are uncommon – but a couple is present on the bottom left flints. This workmanship is thought to be of Victorian origin (see section 4.4.2).	34
Figure 3.13	Evidence of a one-time rampire is obvious in this portion of stretch 'C1' where in the grounds of St Nicholas church the town wall has been removed to enlarge the churchyard. St Nicholas Church viewed here from the north-east and extensively rebuilt after bombing in the last war, stands in the background.	35
Figure 3.14	A Victorian engraving of the hall of the Benedictine Priory which was restored in 1852 and still stands. This engraving was reproduced from Palmer (1864a).	35
Figure 3.15	A boulder of slightly gneissose granite, with dark bands rich in the minerals biotite and hornblende. This broken boulder was set into the external flintwork of the southern end of wall stretch 'C1'. The boulder had, no doubt, originally served as ballast and may have in the first instance been gathered from the Baltic region.	36
Figure 3.16	Set into the flintwork face of wall stretch 'C1', this large igneous boulder was broken in two or more pieces. It had probably initially served as part of a ship's ballast. The rock is a diorite (dolerite) which has subsequently weathered and exfoliated.	36
Figure 3.17	This flint cobble, which is set into the facework of stretch 'C2', shows marked striations on its chalky external cortex. These glacial striae (parallel to the ruler) have been caused by the past movement of the flint within an ice sheet.	36
Figure 3.18	In stretch 'C2' a portion of the town wall has recently been removed to permit more light for St Nicholas School. The bricks of the wall core exposed are mainly of Type 2, but Type 3 bricks are also, present. The top of the Type 3 brick illustrated here reveals a double-margin or sunken margin, a feature which is distinctive but difficult to create or explain.	36
Figure 3.19	The view facing north-east from St Nicholas School of the inside of the wall and its firing bays. These cover the northern part of stretch 'D1' extending into stretch 'C2'. The wall has been capped with modern bricks at the height of the sentry-walk.	37

Figure 3.20 Two small micaceous sandstone seats have been inserted into the walls of this firing bay seen from the grounds of St Nicholas School. Although its outline is poorly visible, the arrow slit in the bay, the sixth to the south of St Nicholas Road, has been bricked up. The same bay is visible in Figure 3.19. 37

Figure 3.21 The view of the exterior of the wall, viewed towards the north-north-west, from Fishers Court, stretch '*D1*'. Here, much of the flint facework has fallen away, to expose a partially repaired, wall core. Type 2 bricks are common in the original core in which the brick to mortar ratio for the wall is unusually high. 38

Figures 3.22 The author holds a tape beside (and to the right of) the line of a small fracture or displacement in part of the core fabric shown in Figure 3.21. This fracture is thought to pre-date the external flintwork, for where this is present it is not displaced. A further example may be observed beside the tape in the figure on the right 38

Figure 3.23 Hospital Tower, viewed from Fishers Court to the north. A squinch arch carries the wall walk to a door in the tower. The ground is noticeably lower on the east side (outside) of the wall. 39

Figure 3.24 Wall stretch '*D2*' and Hospital Tower viewed from the south-east. This external face to the wall is partially enclosed in buildings and in 1884 the wall was completely faced with buildings (in part an abattoir). 'Exotic' rocks are absent in this wall face presumably because it was not exposed over more recent centuries. A new wall has been erected above sentry-walk level.. 40

Figure 3.25 A view southwards along stretch '*E*' of the wall at the northern end of the Market Gates Shopping Centre showing clearly the archwork to the firing bays. 40

Figure 3.26 Detail of the part of the arch of the fourth bay south of Market Gates road partly figured in Figure 3.25. The edge of the arch to the firing bay is constructed of brick headers and stretchers placed alternately in pairs. The bricks are largely of Type 3 and of mid 15th century age. The diameter of the lens cap measures 50mm. 40

Figure 3.27 Guard Tower, set in the Market Gates Shopping precinct and viewed from the south-east............. 41

Figure 3.28 The detail of the underside of the squinch arch visible in Figure 3.27 present on the south side of Guard Tower. From distant observation the bricks used to create the intricate rib-vaulting appear of Type 2, the craftsmanship, therefore, being particularly elaborate for the mid 15th century. 42

Figure 3.29 Wall stretch '*F*', viewed from a point behind No. 4, Alexander Road, very close to the place where Green (1970) described an excavation just outside the wall. The view towards the north, part way along the wall, shows a marked change in the style of the flintwork facing, with the nearer work the more recent (see section *4.4.2*). A visible vent in the wall would suggest a void, probably into a firing bay, but those occupying the building inside the wall know of no access to this basement area. 42

Figure 3.30 Pinnacle Tower as observed from the north shows the remnants of a squinch arch from an original wall walk. The tower has been significantly altered and the flat inside wall completely rebuilt. 43

Figure 3.31 To the rear of the terrace of houses facing on to St Georges Road (stretch '*G*') the East Mount walling is just visible, with access only via the houses. The wall is clearly battered and made of Type 7 bricks with interspersed blocks of ashlar Caen Stone. 44

Figure 3.32 Remarkably, to the west of Ravelin House and St Peter's Plain, stretch '*H*', the outside east face of the wall, has been broken through to gain access to two firing bays (one only shown; Rose (1991, 201) has suggested that these access points were made possibly in the late 18th century when the owners of adjoining cottages broke through the wall to gain an extra cellar). The entrances are now gated and locked. Soil, in the lower levels of the internal rampire, had presumably to be removed to enable their use... 44

Figure 3.33 The outside of the town wall, stretch 'H', behind the mews to the west of St Peter's Plain. A putlock hole is visible in the flint facework. To the left, the face has fallen to expose the wall core. 45

Figure 3.34 The stump of Harris's Tower, surmounted by an early 19th century house, as viewed from the north. The wall changes slightly in direction at this point. Palmer (1864a, 115) indicated that the ground floor of the tower had been used as a stable. 45

Figure 3.35 A view from the north-west of the inside of the wall in stretch '*J*' on Dene Side to the north of the Community Services building. Evidence of the rampire remains, and it still rises to almost cover the firing bays. Above the wall walk, which has been repaired with flint beach cobbles, the battlements have been extensively altered. 46

Figure 3.36 A similar view from the north-west of the inside of the wall to the south of the Community Services building in Dene Side (stretch '*J*'). The firing bays become increasingly exposed proceeding towards the south and Garden Gate. In nearly all instances the bays have been removed to leave only their traces on the inside of the wall. The remnant bases of the firing

	bay piers remain (see also Figures 3.37 and 3.38).	47
Figure 3.37	The trace of the 4th firing bay arch to the north of Garden Gate. The bricks outlining the Gothic style arch are all of Type 2 bricks and the arch is probably of original mid 14th century construction. Inside the arch the wall has been rebuilt and it is of a later date.	48
Figure 3.38	Slightly more of this firing bay, the 12th bay north of Garden Gate, is preserved. Again, the arrow slit has been blocked. Most of the bricks are again of Type 2.	48
Figure 3.39	The arrow slit of the 4th firing bay north of Garden Gate viewed from the outside of the wall (compare with Figure 3.37). The slit has been completely replaced and infilled with bricks. These are displayed as headers of mainly Types 3, 6 and 7 bricks, but there are also a few of Type 4, and possibly at least one modern brick.	48
Figure 3.40	This figure displays the next arrow slit (the fifth) north of that shown in Figure 3.39. Set differently, but again completely replacing the original slit, this contains only various old bricks, and the replacement may have been done at the time of the rampiring in the mid 16th century.	48
Figure 3.41	At least three different styles of flint facework can be observed in this wall stretch 'J', opposite No. 7, Blackfriars' Road, near the Time and Tide Museum. Some of the workmanship is considered to be no older than Victorian.	49
Figure 3.42	A view of the wall towards the north from Garden Gate. Blackfriars' Road runs parallel to the wall on the right. The sentry-walk is preserved above the first few firing bays. The south, near end of the wall has been rebuilt and includes a Caen Stone cross presumably created from reused stone.	49
Figure 3.43	The external face of the wall in stretch 'K', slightly to the south of the pottery. The moderately squared flints to the south (left) of this view are probably the oldest in this stretch of wall face. Where the flint facework has fallen the material in view has also been repaired, that is, it is not original wall core: note that it contains flint beach cobbles.	50
Figure 3.44	This exposed wall in which all trace of the flint facework has been lost, occurs outside the pottery in stretch 'K'. On first impression the lowest third (2m.) of the wall with numerous fragmentary bricks appears old, but the bricks are of different types and the wall contains abundant rounded beach cobbles. Higher, bricks are far less numerous and a chalky mortar forms much of the wall matrix. This probably represents the early wall core with bricks no younger than the mid 16th century.	50
Figure 3.45	Great Yarmouth's most imposing tower is probably the South East Tower in Blackfriars' Road. Here the tower is viewed for the south-east. The four rows of chequerwork, although, like the tower, partially repaired, are believed to date from the 16th century and in this instance most of the visible, rounded beach cobbles are likely to be of this age. The wall break, created post 1904, is noticeable immediately this side of the tower.	51
Figure 3.46	A view very similar to that shown in Figure 3.45 is portrayed in this pen and ink drawing by Noel Spencer. The drawing is exhibited in the Time and Tide Museum and was an accession to the Great Yarmouth collections in 1983. Permission to reproduce this drawing is gratefully acknowledged.	51
Figure 3.47	This etching of the South East Tower from the north-west was executed by Mrs Bowyer Vaux and appeared in Palmer (1852, where it was incorrectly labelled the 'south west tower'). Considerable alterations to this view have occurred in the last 150 years.	52
Figure 3.48	Palmer (1864a, 114) referred to three stone 'gurgoyles' on the walls of the South East Tower. High on the walls, only one gargoyle, probably of Caen Stone, remains. It seems likely that this was reused from the ruined Blackfriars' Church.	52
Figure 3.49	This gun port on the South East Tower is constructed of 'roach' rock; a variety of Upper Jurassic, Portland Stone. High in the tower, similar gun ports are made of Caen Stone, for which, in the lower ports, the Portland Stone is probably a replacement.	52
Figure 3.50	The external surface of wall stretch 'L1'. This wall displays numerous reused blocks of stone from the earlier Blackfriars' site. These include quatrefoil column sections (top left) probably from the church, as well as window-framing stones now reversed to splay outwards as embrasures. The height of these embrasures gives a realistic idea of the original size of the internal rampire upon which any defensive canons would stand.	53
Figure 3.51	Tightly packed squared flints with gallets, on the outer surface of part of wall stretch 'L1'. The different styles of flint facework are discussed in section *4.4.2*. It follows, that if this is apparently the earliest flintwork style in the wall, which here dates from the 1550s, the style is also likely to have been created in the mid 16th century.	53
Figure 3.52	This block of Barnack Stone (centre) is darker on the right where it has been discoloured red by fire. Set in this stretch 'L1' of the outside of the wall, it is enclosed by three pieces of Caen Stone and flints with galleting. Both the Caen and the Barnack Stones have apparently been reused from the fire devastated ruins of the Blackfriars' monastery and church, when the wall	

	was rebuilt in the mid 16th century.	54
Figure 3.53	An off-set occurs in the wall between stretches '*L1*' and '*L2*' and it is shown here viewed from the south-east. The south end of wall '*L1*' terminates in a Caen Stone quoin constructed at other than a right angle, suggesting that it was cut for the purpose, and is not, therefore, reused.	54
Figure 3.54	This large, rounded cobble of Jurassic muddy limestone occurs within the outer flint face of the wall in stretch '*L2*'. The cobble shows evidence of its previous existence in a shallow marine environment in that it shows borings (here minute holes) by the marine organism *Polydora*. This suggests that the cobble was earlier used as ballast. The lens cap has a diameter of 50mm.	55
Figure 3.55	The wall is seen here in section at the Charles Street pathway, southern end, of stretch '*L2*'. The flint wall face, in this case repaired, can be seen to be of limited thickness. The external face of the wall can also be seen to include numerous ashlar blocks from the earlier Blackfriars' site.	55
Figure 3.56	Charles Street pathway viewed from the south-east with wall stretch '*L3*' in the left foreground. An unusual wall break, created with mid 16th century bricks is evident in the wall. To the south of this break the wall has lost its flint facing.	55
Figure 3.57	The wall break viewed in Figure 3.56 is seen here in more detail. The wall to the south (left) has lost the external flint facework to expose the wall core.	55
Figure 3.58	Inside the wall stretch '*L3*', just north of Blackfriars' Tower, exotic boulders and cobbles are common. That the wall has been much repaired is evident from the many rounded flint beach cobbles and the row of igneous and much altered sedimentary rocks, some of which may be altered tuffs, were almost certainly previously used as ships' ballast.	56
Figure 3.59	The wall core is exposed occasionally on the inside of wall stretch '*L3*'. Flints and mortar both exceed brick fragments in quantity and the wall also includes some Chalk fragments which can be seen in this figure. The fragmentary bricks are almost exclusively of Types 6 and 7. The steel ruler is 300mm. in length.	56
Figure 3.60	Blackfriars' Tower is here viewed from the south-east. It supports three chequerwork rows of ornamentation. The passage through the tower was made in 1807. Blackfriars' Tower, as other towers, has been extensively repaired over a variety of different periods.	57
Figure 3.61	Noel Spencer's pen and ink drawing reveals the extent of the alterations to buildings surrounding the tower over the last half century. The view is similar to that in Figure 3.60. The drawing, a Yarmouth Museums' accession in 1983, is displayed in the Time and Tide Museum, and to which permission to publish is gratefully acknowledged.	57
Figure 3.62	This figure of the rear of Blackfriars' Tower, taken from the north, was first published by Tingey (1913, where it was incorrectly described as the South East Tower). The changes to the adjoining walls and their relationships to enclosing buildings in less than a century are dramatic.	57
Figure 3.63	Wall stretch '*M2*' is seen here from the east. The inside of the wall is greatly defaced; the firing bays having been removed and their presence can only be detected by the positions of the much altered arrow slits. This wall, over recent centuries, was enclosed by attached buildings and, it seems possible that the firing bays were removed at the time of rampiring in the 16th century.	58
Figure 3.64	This figure displays the outside of the wall shown in Figure 3.63, observed from the south-east. Although the evidence of rampiring internally is slight, the difference in ground levels on either side of the wall is obvious. The wall shows evidence of repairs, and particularly of those undertaken in the last century.	58
Figure 3.65	Palmer's Tower is here seen from the south-east. The wall projecting from it towards the south (left) is modern. Again, much altered, the tower may have once supported a windmill.	58
Figure 3.66	Noel Spencer's pen and ink drawing of Palmer's Tower, again kindly reproduced from the Time and Tide Museum's, 1983 acquisition, collection, is a view from slightly more to the south than that shown in Figure 3.65.	58
Figure 3.67	This etching of the outside of the South Gate was prepared for Dawson Turner, the East Anglian 19th century historian. Although the North and South Gates to the town were the same width (a gate of 3.7m. set in a total width of 20.1m.; Swinden, 1772), they were somewhat different in style (compare with Figure 3.6). The figure is reproduced from Preston (1819).	59
Figure 4.1	The distribution of principal brick types around the Great Yarmouth town wall, mostly related to the presence of wall core exposures (see also Tables 4.1 and 4.2). At locality '*B3*' access and detailed measurements were not possible (na = not applicable). 'Modern' bricks have been excluded from these analyses.	62
Figure 4.2	The wall to the east of St Nicholas School (stretch '*C2*') has in relatively recent years been	

	broken through to provide light for the classrooms. This view, from inside the wall and towards the north, illustrates the many Type 2 bricks visible in the wall core. One Type 3 thinner brick can also be seen. The flint wall face is on the right.	63
Figure 4.3	Faden's plan of Great Yarmouth dating from 1797; the map was published in *The Archaeological Journal*, 1980. This copy has been reproduced by kind permission of the Norfolk Records Office. Cobholm Island is marked at the western edge of the plan on the south side of Breydon Water. Note also the moat or ditch well outside the walls to the north of the town.	64
Figure 4.4	The detail of a reused, weathered and broken slab of *Viviparus* limestone ('Purbeck Marble') observed in the flint facework of wall stretch '*L1*'. The numerous remains of the fresh-water gastropod *Viviparus* can be seen in the rock.	67
Figure 4.5	Part of the external flintwork to the wall stretch '*J*' (see section *3.2.8*); this area being towards the northern end of the wall length near to Time and Tide Museum (approximately opposite No. 7, Blackfriars' Road). This style of flintwork appears to date from prior to the mid 16th century and is possibly the oldest style in the wall stretch (style 2). Notice the purposefully broken flints are well coursed and reasonably widely spaced (see also Figure 4.11).	70
Figure 4.6	A view of a different part of the same wall stretch '*J*' as Figure 4.5 (approximately opposite No. 16, Blackfriars' Road). Here the flints are tightly packed but remain moderately well coursed. Packing, to present an almost continuous flint surface, is enhanced with flint galleting. This pattern is typical of style 3 and is thought to indicate a post 1550s age (see also Figure 4.12).	70
Figure 4.7	A variety of style 3 craftsmanship that is much less common than that seen in Figure 4.6 is here illustrated from the east face of the North West Tower. The flints are moderately well squared so that they fit closely and gallets are only rarely used.	70
Figure 4.8	Nipples are absent from the broken surfaces of these flints. This area of flintwork again occurs in the wall stretch '*J*' opposite the Time and Tide Museum (approximately opposite No. 9, Blackfriars' Road). The workmanship is moderately coursed in this instance but elsewhere in the same area the flints are randomly related to each other, typical of Victorian work in the region of style 4 (see also Figure 4.14).	70
Figure 4.9	A further variety of Victorian (or more recent) flintwork is seen in this area of stretch '*J*' wall (approximately opposite No. 8 Blackfriars' Road). In this instance 'half' bricks have been randomly added to the wall for supplementary ornamentation. The relationships and features of the broken flint cobbles are otherwise not unlike the style 4 (Figure 4.8) workmanship.	71
Figure 4.10	The battlements on the top of the stretch '*J*' wall have been the most recently repaired. Unbroken cobbles of beach flints, well coursed and in this instance interspersed with mainly modern bricks, are typical of style 5 of the flint facework patterns.	71
Figure 4.11	Detail of style 2 flints in the area of Figure 4.5 of stretch '*J*' of the Great Yarmouth town wall. Nipples, and opposing cups, are abundant on the broken exposed surfaces of the flints. Unbroken surfaces of many of the flint cobbles show curvature resulting from beach erosion. The lens cap has a diameter of 50mm. This figure should be compared with style 3 faces seen in Figures 4.12 and 4.13.	71
Figure 4.12	The detail of a style 3 flintwork example from the area of Figure 4.6 of wall stretch '*J*'. Nipples and cups are fairly common, but the flints have been knapped to provide better fit and gallets built-in to fill any gaps. The lens cap is 50mm. in diameter, and the figure should be compared with style 2 work seen in Figure 4.11.	71
Figure 4.13	A further example of style 3 flintwork. This forms a very small patch in the area of Figure 4.6 and is more similar to the style seen in Figure 4.7. The lens cap is 50mm. in diameter.	71
Figure 4.14	The detail of the style 4 flintwork from the area of Figure 4.8 in wall stretch 'J'. Note the absence of nipples on the broken flint faces.	71
Figure 5.1	An engraving of the South East Tower as it appeared in *The Builder*, 1886. The tower is viewed from the south-east and clearly displays its chequerwork.	74
Figure 5.2	An engraving of Blackfriars Tower reproduced from *The Builder*, 1886. Both this engraving and that in Figure 5.1 were completed by Mr G. Ashburner and they appeared on page 357. The tower is viewed from the east.	74
Figure 5.3	A copy of the picture-map of Great Yarmouth reproduced by Palmer (1854). The original copy of the plan which is thought to date from about 1585, is often referred to as the Cottonian map. It is held in the British Library. Reproduced by kind permission of Norfolk Records Office, Norwich.	75
Figure 5.4	The rampire to the north side of the graveyard at St Nicholas Church viewed from the west. All possible firing bays on the inside of the town wall have been buried beneath the earth embankment which rises to the wall from the south.	76
Figure 5.5	Areas of the Great Yarmouth town wall which indicate evidence or limited evidence of the	

	presence of the 16th century rampiring.	77
Figure 5.6	A plan of the East Mount, Great Yarmouth created by Rose (1991) in 1984 at the time of building demolitions and excavations at the site. The plan is reproduced by the kind permission of the author and the Norfolk and Norwich Archaeological Society. Of the inset plans, Faden, 1797, is reproduced as Figure 4.3 within the present work. Figure 3.30 was taken to the rear of Nos 89 and 88, St. George's Road towards the south-east.	78
Figure 5.7	A more greatly enlarged example of the Hatfield House picture-map which appears as the frontispiece to the present work. The plan, here displaying most of the town walls (north to the right), is thought to date from about 1588 and have been prepared in preparation for the possible invasion of the Spanish Armada. The plan depicts the East Mount (K) as a true ravelin, although there is no evidence that it was ever built in this form. Note also the absence of any moat external to the walls and King Henry's Tower (F) being illustrated as square. The plan is reproduced by kind permission of the Norfolk Record Office.	79
Figure 5.8	The plan of Great Yarmouth prepared by Sir Robert Paston in 1688. This copy was reproduced from Rutledge and Rutledge (1978). The plan is held in the Norfolk and Norwich Archaeological Society Library. Permission to publish is gratefully acknowledged.	80
Figure 5.9	This plan was constructed by Mr A. W. Morant for Palmer (1864a) and is supposed to illustrate Great Yarmouth about 1650. This appears to be the only plan which shows a moat adjoining and completely encircling the town walls. It shows the later moat, possibly of 1642, running into it from the north. The plan is interpretive of Palmer's personal views regarding the presence of a moat.	83
Figure 5.10	A plan originally thought to date from 1619 (Ecclestone and Ecclestone, 1959) and reproduced here from that privately printed volume. It is held in the British Museum. It is now thought to have been prepared by R. Cory and to date from about 1820 and be based on 17th century information. The plan clearly displays the moat at the north end of the town and well outside the walls.	84
Figure 8.1	This pen and ink drawing of the North West Tower viewed from the River Bure to the west, was completed about 1955 by Noel Spencer. It is displayed in the Time and Tide Museum from where it was kindly reproduced.	97
Figure 8.2	The South Gate: an etching undertaken in 1812 by John Sell Cotman. In that year the gate is said to have been demolished. Compare this figure with Figure 3.67 published in 1819. This figure was reproduced by kind permission of the Norwich Castle Museum and Art Gallery.	98
Figure 8.3	This watercolour of the rear of the South East Tower was also completed by Cotman in 1812. It is again kindly reproduced from the Norwich Castle Museum and Art Gallery collections. In 1812 the rampire had suffered little destruction.	99
Figure 8.4	The River Bure, the North West Tower and the old spire of St Nicholas Church, all figure in this delightful scene painted by W. H. Hunt in watercolour, in 1860. This figure was copied, with kind permission, from the original which hangs in the Time and Tide Museum, Great Yarmouth.	99

A LIST OF INCLUDED OLD MAPS AND PLANS BY DATE ORDER

Figure 1.3   ROMAN – Modern interpretation
A reconstruction of the Norfolk coastline in Roman times. The approximate line of the modern coastline is indicated by a broken line and the position of Great Yarmouth by an asterisk. Some of the inland water area would have been marshlands. (Partly after Pearson, 2002)... 3

Figure 1.2   ANGLO-SAXON – Elizabethan interpretation
This early map, normally referred to as the 'Hutch Map', has been attributed to the Elizabethan period (Ecclestone and Ecclestone, 1959, 20) and is thought to depict the immediate Norfolk coastline during Anglo-Saxon times. Interpretation is made more complex by geographical north being placed at the bottom of the original map. Reproduced from Ecclestone and Ecclestone (1959). The original is now held in the Norfolk Record Office (Y/C 37/1).. 2

Figure 1.5   LATE 13th CENTURY – Modern interpretation
A plan of Great Yarmouth in the late 13th century, copied by kind permission from Rutledge (1990). 6

Figure 5.3   ABOUT 1585
A copy of the picture-map of Great Yarmouth reproduced by Palmer (1854). The original copy of the plan which is thought to date from about 1585, is often referred to as the Cottonian map. It is held in the British Library. Reproduced by kind permission of Norfolk Records Office, Norwich. 75

Frontispiece   ABOUT 1588
A plan of the Great Yarmouth fortifications as they are supposed to have appeared in 1588 and compiled at about that time. The plan is normally described as the Hatfield House map and the original is held by the British Museum (O'Neil and Stephens, 1942). The plan is reproduced here by kind permission from a copy held by the Norfolk Record Office. As far as can be interpreted the key reads: A – West Gate, B – North Tower, C – The North Ravely(i)n, D – Nor(th) East Tower, E – Corner Tower, F – K. Henry's Tower, G – St Nicholas Tower, H – Prieery Tower, J – Market Gate, K – the Ra(velin), L – (H)ospitall Tower, M – Bennets Tower, N – Oxne Tower, O – Pinakell Tower, P – Shanes Tower, Q – mount and curtyns, R – the mountes bulwark, S – Bowls. Tower, T – East Tower, V – New wall, W – Sow Est (Tower), X – Myddell Tow(er), Y – Sowth Gate. These names are not necessarily the same as those given to the corresponding structures by subsequent authors or within the present work Part of this plan is also reproduced as Figure 5.6. ....Frontispiece

Figure 5.7   ABOUT 1588
A more greatly enlarged example of the Hatfield House picture-map which appears as the frontispiece to the present work. The plan, here displaying most of the town walls (north to the right), is thought to date from about 1588 and have been prepared in preparation for the possible invasion of the Spanish Armada. The plan depicts the East Mount (K) as a true ravelin, although there is no evidence that it was ever built in this form. Note also the absence of any moat external to the walls and King Henry's Tower (F) being illustrated as square. The plan is reproduced by kind permission of the Norfolk Record Office. 79

Figure 5.10   ABOUT 1619 (1820 interpretation)
A plan originally thought to date from 1619 (Ecclestone and Ecclestone, 1959) and reproduced here from that privately printed volume. It is held in the British Museum. It is now thought to have been prepared by R. Cory and to date from about 1820 and be based on 17th century information. The plan clearly displays the moat at the north end of the town and well outside the walls. 84

Figure 5.9   ABOUT 1650 (1864 interpretation)
This plan was constructed by Mr A. W. Morant for Palmer (1864a) and is supposed to

	illustrate Great Yarmouth about 1650. This appears to be the only plan which shows a moat adjoining and completely encircling the town walls. It shows the later moat, possibly of 1642, running into it from the north. The plan is interpretive of Palmer's personal views regarding the presence of a moat.	83
Figure 5.8	1688 The plan of Great Yarmouth prepared by Sir Robert Paston in 1688. This copy was reproduced from Rutledge and Rutledge (1978). The plan is held in the Norfolk and Norwich Archaeological Society Library. Permission to publish is gratefully acknowledged.	80
Figure 3.2	ABOUT 1750 Part of Swinden's Plan of Great Yarmouth produced to accompany his work of 1772. This copy has been reproduced, with their kind permission, from The Society of Antiquaries of London (1953). The map is thought to date from 1758, rather than 1738 as proposed by O'Neil (1953).	28
Figure 4.3	1797 Faden's plan of Great Yarmouth dating from 1797; the map was published in *The Archaeological Journal*, 1980. This copy has been reproduced by kind permission of the Norfolk Records Office. Cobholm Island is marked at the western edge of the plan on the south side of Breydon Water. Note also the moat or ditch well outside the walls to the north of the town.	64

ACKNOWLEDGEMENTS

IMPORTANT STATEMENT

The present work presents what would appear to be the first multidisciplinary study of the Great Yarmouth walls. The work was initiated by Dr Peter Hoare[1] who, following successful joint studies of the town walls of both King's Lynn and Sandwich, suggested that a similar examination of the Great Yarmouth Town Wall was both necessary and likely to be productive. Following a brief review of the existing literature and an inspection of some of the walls, Dr Hoare invited the present author to join him in the overall study. Although, thanks to a reasonably prolific amount of early documentation, archaeological and more especially historic elements of the walls' construction had been examined, a detailed assessment of the walls' fabric and its geology had not been attempted. Early in this assessment it became clear that, with much of the wall composition consisting of a wide variety of bricks, a national expert on brick identification, Peter Minter[2], should be invited to identify and confirm various brick types, their possible origins and dates. This support has proved particularly valuable in drawing conclusions as to the ages of reconstruction of the various walls. Subsequently, Paul Rutledge[3] was invited to assist with this work and his experience in providing information on the early manuscripts and documentation has also proved invaluable.

For a wide variety of reasons, but probably much to do with the complexities of multidisciplinary analyses, the present author has been left to consolidate the writings and verbal contributions of the other three persons into those of his own work. The present author has tried to check the accuracy and veracity of each of these observations. Any errors which may be present in this publication are, therefore, the responsibility of the author.

1. Dr Peter G. Hoare, lives in Ely, but may be contacted at The School of Geosciences, The University of Sydney, Sydney. New South Wales 2006, Australia.
2. Mr Peter Minter, Managing Director, The Bulmer Brick and Tile Company, The Brickfields, Hedingham Road, Bulmer, Sudbury. Suffolk CO10 7EF. Peter Minter's expertise relates in part to his Company specialising in the manufacture of replacement early Roman and historic bricks.
3. Mr Paul Rutledge, 'The Pleasance', 4, Queen Street, New Buckenham. Norfolk NR16 2AL. Paul Rutledge has published extensively on early historic Great Yarmouth.

The author must firstly offer his prolific thanks to Dr Peter Hoare, and Messrs Peter Minter and Paul Rutledge (see 'Important Statement' above). Without their support this work would not have been completed. The publishers of British Archaeological Reports did much to simplify and ease the progress of this publication and their assistance in this way is greatly appreciated. Great Yarmouth Borough Council (Historic Buildings Budget) kindly provided a grant which assisted in defraying the cost of publishing some of the figures in this work.

The assistance of numerous library and museum staff must also be acknowledged; especially those at Fleet Library, Hampshire. In particular, I would wish to mention Darren Barker, Conservation Officer, Great Yarmouth Borough Council; Fiona Ford, Maritime Curator, Great Yarmouth Museums; Vanessa Trevelyan, Head of Museums and Archaeology, Shirehall, Norwich; Norma Watt, Norwich Castle Museum and Art Gallery; the staff in the archive section of the Norfolk Record Office and those who kindly provided assistance at the Time and Tide Museum, Great Yarmouth. My thanks are due also to Stephen Hart for a useful correspondence on the early use of flints.

Finally, the author wishes to express his thanks to his patient wife and his family and friends who put up with both his absence and demands over the long periods of researching and compiling this work.

IMPORTANT NOTICE

PROCEDURES

Following initial field visits to the Great Yarmouth town walls it became evident that in structure and composition they were extremely variable. Each portion of visible wall revealed a new range of worked materials and styles of building. This applied particularly to the external surfaces of the walls where extensive patching was the norm. It proved necessary to fully comprehend the extent of these variations prior to any systematic study. A leading national expert on medieval bricks, Peter Minter, was invited to assist in the identification and confirmation of the many brick types. The gradual assimilation of information concerning the total picture regarding the wall materials present and the patterns in which they had been employed involved frequent return visits to check previously determined minutiae against the details provided from newly viewed sites of the walls. Finally, with a composite picture of most of the materials and building styles involved it proved possible to undertake a full and continuous examination of the full circuit of the walls.

The text of this work, therefore, follows the procedures adopted in the field. Following an historic preamble in Chapter 1; Chapter 2 describes the broad conclusions reached concerning the materials used in the fabric of the walls and the building styles adopted in their use. Chapter 3 then examines the town wall circuit in some detail. It proved far too cumbersome and incomprehensible to discuss the different fabric varieties and styles without initially introducing them in Chapter 2. An introduction to the characteristics of each new brick type as it was first encountered, for instance, would have been most confusing. Those wishing to understand the broader aspects of this work may, therefore, omit the detail of Chapter 3 and progress to the full fabric analyses and conclusions provided in Chapter 4. It should be stated that the descriptions provided in Chapter 3, for the sake of brevity, often only summarize the visual evidence.

CHAPTER 1

GREAT YARMOUTH: GEOGRAPHICAL, HISTORICAL AND ECONOMIC BACKGROUND

*'Few towns can boast a history so romantic, so
fascinating and so unusual as that of
Great Yarmouth'*
Ecclestone and Ecclestone (1959)

1.1 Introduction

Great Yarmouth, located on the south-east coast of Norfolk (TG 5207; 52° 36.4'N, 1° 43.7'E) (Figure 1.1), possesses one of the most complete medieval town walls in England. The full circuit exceeded 2 km. in length and contained ten gates and 18 towers (Swinden, 1772, 82, 96-101). Nearly two-thirds of the walls are still standing, including 11 towers, a number of them largely intact. Considerable stretches of wall retain their sentry-walk, arched firing bays and battlements. The wall is one of a small number in England where associated pre-Civil War earthworks may still be seen.

Previous accounts of Great Yarmouth's town wall, towers and gates underestimated their geological richness and provided inadequate statements on the provenance of the construction materials. The aims of the present study are (i) to summarise the state of the existing knowledge related to the town walls, (ii) to record in detail the geological content of the walls, (iii) to endeavour to trace the ultimate sources of those constituents which comprise the main body of the wall, and (iv) to consider the geology and structure of individual stretches of wall and by so doing establish the periods and order in which they were built, repaired and reconstructed.

1.2 The foundation and development of Great Yarmouth

1.2.1 Site and situation

Great Yarmouth is built on land which once formed a sand bank close to the confluence of the Rivers Bure, Yare and Waveney. An early, Elizabethan map (Figure 1.2), generally attributed to Thomas Damet and known as the 'Hutch Map', is drawn with geographical north at the foot of the map, and depicts this relationship. The map is unusually named in recognition of its original storage place, in a town chest known as 'The Hutch' (Palmer, 1864b). In the Roman period (Figure 1.3), a wide estuary existed in the region, with the military fort of Caister-on-Sea (TG 516 123) to the north and Burgh Castle (TG 475 046), a 'Saxon' shore fort, in the south (Wallis, 1995).

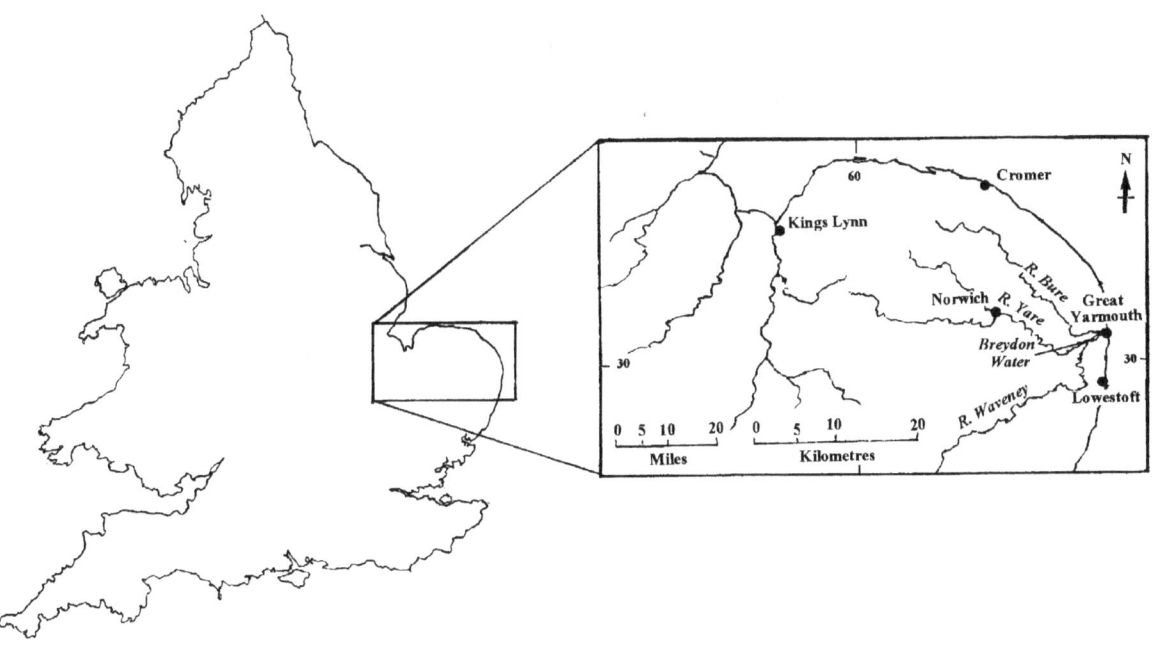

Figure 1.1 The geographical situation of Great Yarmouth. The town is situated in the National Grid square TG.

The Medieval Town Wall of Great Yarmouth, Norfolk, U.K.

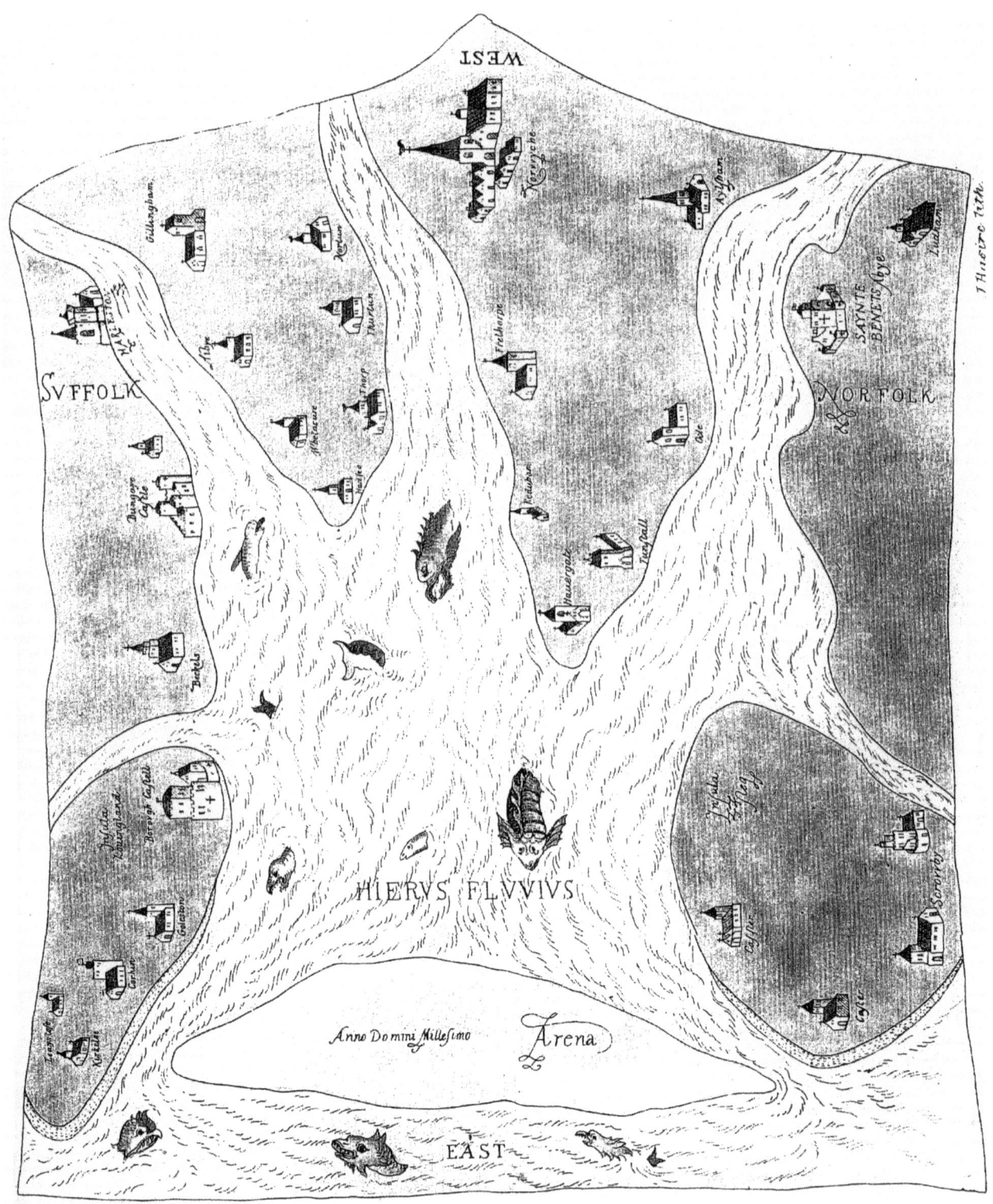

Figure 1.2 This early map, normally referred to as the 'Hutch Map', has been attributed to the Elizabethan period (Ecclestone and Ecclestone, 1959, 20) and is thought to depict the immediate Norfolk coastline during Anglo-Saxon times. Interpretation is made more complex by geographical north being placed at the bottom of the original map (now placed on the right). Reproduced from Ecclestone and Ecclestone (1959). The original is now held in the Norfolk Record Office (Y/C 37/1).

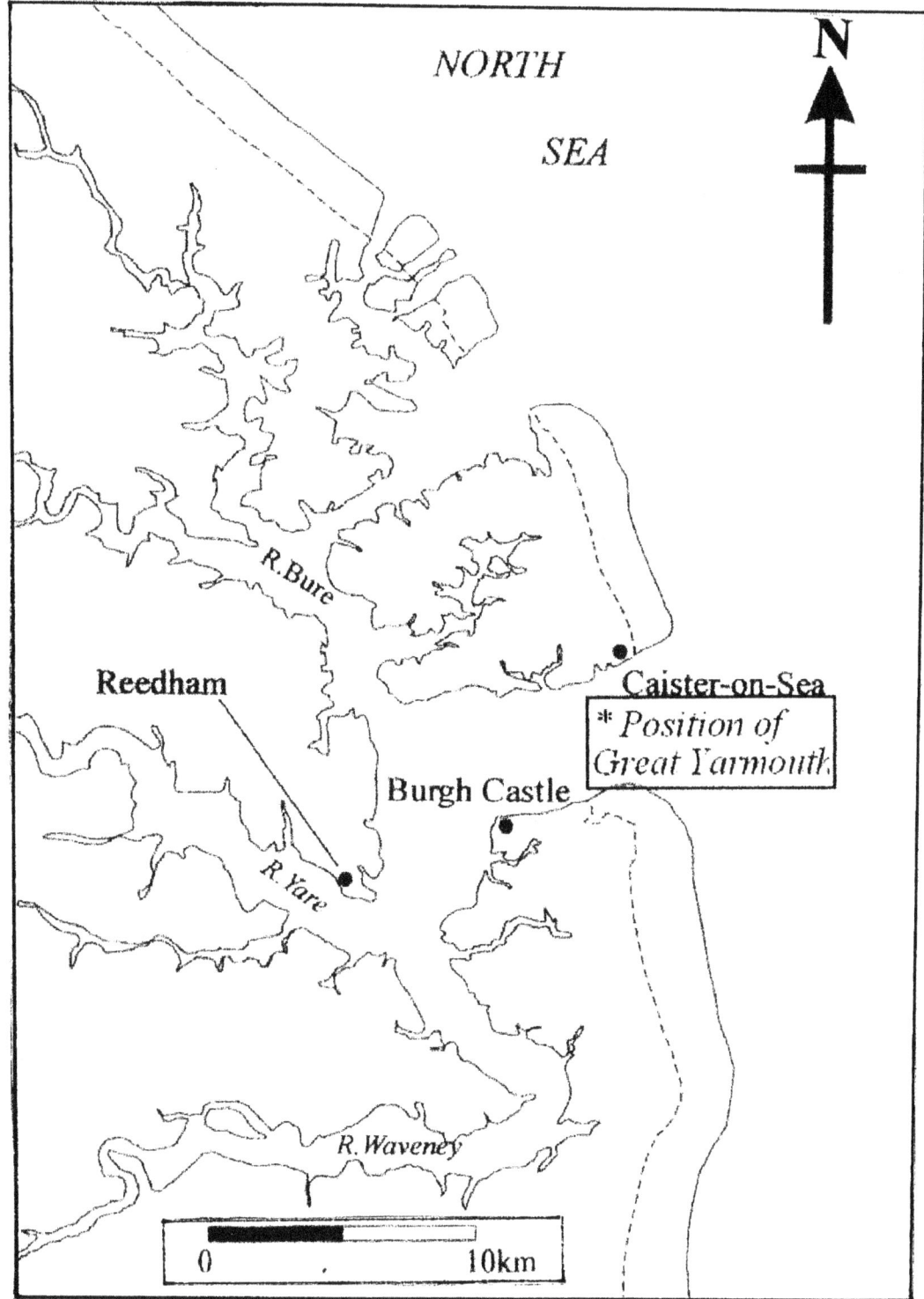

Figure 1.3 A reconstruction of the Norfolk coastline in Roman times. The approximate line of the modern coastline is indicated by a broken line and the position of Great Yarmouth by an asterisk. Some of the inland water area would have been marshlands. (Partly after Pearson, 2002).

Both Roman localities having equal claim to the title *Gariannonum* (Pearson, 2002, 15). The sand bank upon which Great Yarmouth now stands probably became linked to the mainland in the north by a sand spit (today forming the sands and gravels of the North Denes Formation) as little as a millennium ago. The sand spit may not have extended southwards to its fullest until about 1200 (Emery, 1998), but then reaching as far as Corton (TM 54 97). The continual movement of sand from north to south along the coast has played an important role in the fortunes of Great Yarmouth, repeatedly closing the access to the river mouths and necessitating the repeated building of artificial navigation channels known as havens. The construction of that in use today, the seventh haven, was commenced in 1560 (Manship, 1619, 90), first cut in 1566, and apparently

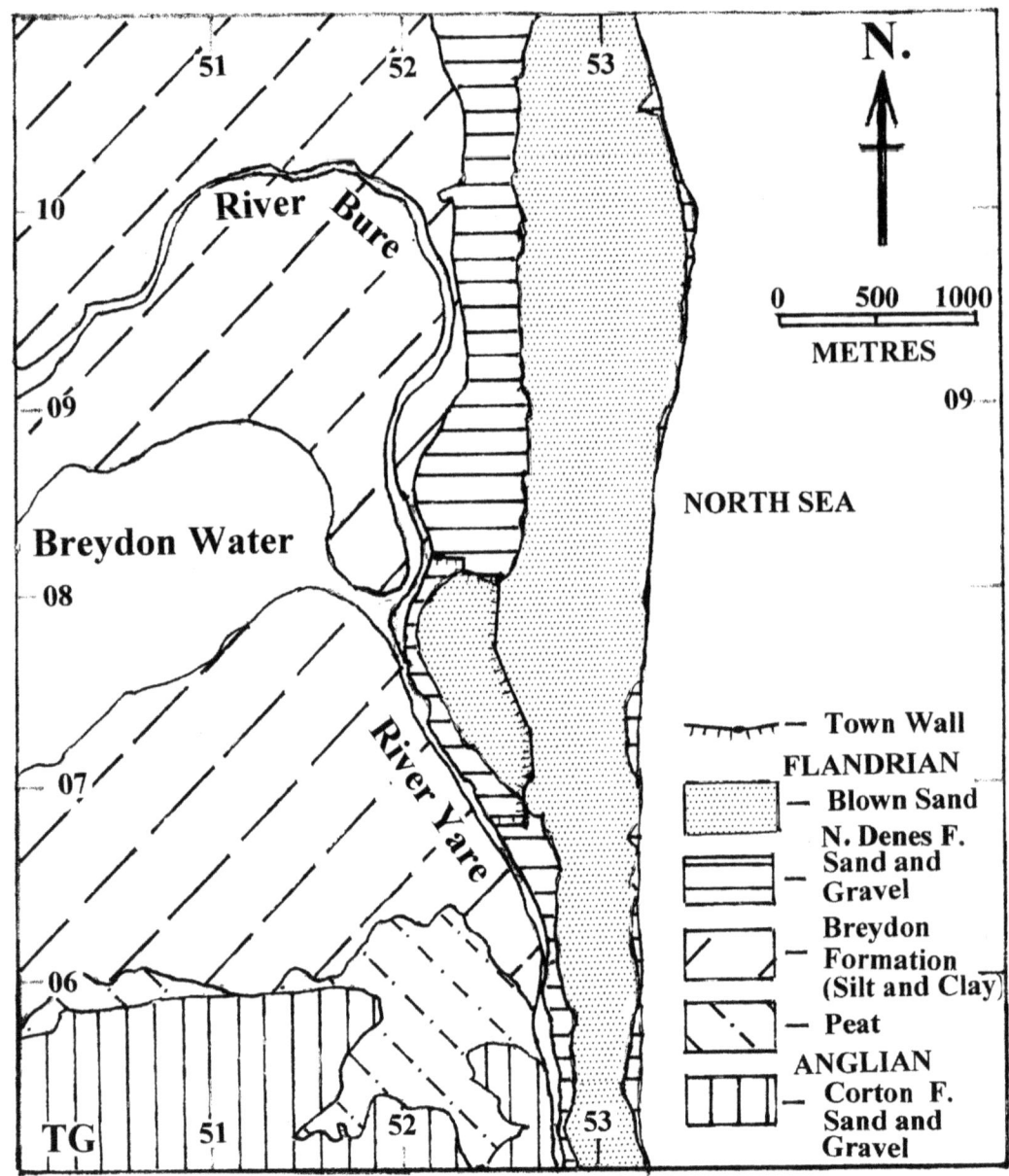

Figure 1.4 The Quaternary geology in the immediate vicinity of Great Yarmouth. Holocene, Flandrian deposits (N. Denes F. = North Denes Formation) rest upon Pleistocene, Anglian deposits (Corton F. = Corton Formation).

completed in 1613 (Pevsner and Wilson, 1997), opposite Gorleston-on-Sea.

At depth below Yarmouth, a late Pliocene-early Pleistocene sequence rests on Upper Cretaceous Chalk Group strata at -155m. O.D. beneath the town (Arthurton, *et al.*, 1994, figure 15). The Superficial geology of the area includes patches of 'Glacial sand and gravel' and till, but it is dominated by Holocene alluvium (Arthurton, *et al.*, 1994). The immediate deposits below the walled town, however, are of Holcene, Flandrian 'Blown sand', but near to the Rivers Yare and Bure this gives way to underlying Flandrian, North Denes Formation 'Sand and gravel' (*Gt Yarmouth, Sheet 162, Quaternary and Pre-Quaternary Geology*, 1: 50,000 Map, 1990) (Figure 1.4). Relative sea level at Great Yarmouth in the thirteenth century probably stood between Ordnance Datum and +1m. O.D. (Arthurton, *et al.*,1994, figure 55); *contra* Green and Hutchinson (1960, 114; 1965, 86) and Green (1970, 114).

It has been suggested that Yarmouth did not officially become 'Great' until Henry III's charter of 1272 (Palmer, 1854, 198). However, the settlement on the other (western) side of the River Yare was referred to as Little Yarmouth as early as 1131 (Page, 1906, 437), the implication being that, informally, Great Yarmouth must have existed from at least that time. The Domesday Book places Yarmouth's origins before 1066 (Brown, 1984, I, 67; Rutledge, 1999, 27). The name Yarmouth [*Gernemwa* or *Gernemutha*] first appears in 1086 (Sandred *et al.*, 1996, 27) as part of the Hundred of Est Flec [East Flegg]

(Page, 1906, 47), and Great Yarmouth [*Magna Gernemue*] in 1252 (Sandred *et al.*, 1996, 28).

'Medieval Great Yarmouth consisted of an unusual gridded street pattern in which buildings fronted onto three main streets aligned north-to-south. In turn these buildings were separated by a hundred and fifty alleyways – known locally as Rows – running from the market place westwards to the quays of the River Yare. This pattern may reflect the town's origins in the eleventh century as a new trading post under Anglo-Scandinavian control; similar street patterns can be observed in Scandinavian towns of the same period.
The Norfolk Archaeology Unit Annual Review, 1997-1998

1.2.2 Brief economic background

The Roman sites of Caister-on-Sea, about 5km. to the north of Great Yarmouth and Burgh Castle some 5km. to the south-west (Figure 1.3), were probably abandoned as subsequent settlement became focused on the site of the new town. A description of the topography of the town early in its history is provided by Rutledge and Rutledge (1978) and Rutledge (1990, from which see the current Figure 1.5). Great Yarmouth was already a borough by the time of the Domesday records in 1086 (Rutledge, 1976, 8), although permanent coastal settlement may not have been possible before the Danish raids ended in the 11th century. The Domesday Book records that 'King Edward held Yarmouth. Always 70 burgesses' (Brown, 1984, folio 118b), and that '... 24 fishermen belong to this [Gorleston?] manor' (Rumble, 1986, folio 283a). This seems to imply Yarmouth's origin as a seasonal fishermen's camp (compare Somerton and Winterton just to the north).

Great Yarmouth grew quickly. When it received its first royal charter in 1208 (Manship, 1619, 2) its fee-farm or yearly rent to the Crown was approximately half that of the Norwich. The town apparently reached its medieval economic peak between 1209 and 1336, largely on the basis of its position as England's major source of herring (Saul, 1979, 105; 1981, 33; Carter, 1980a, 300). Ready access to the Low Countries and to the Baltic, as well as to the river system extending into the English interior, resulted by the early 1330s, in it surpassing in wealth the larger and longer-established Norwich (Hudson, 1907). Of the English provincial towns, only Bristol, York and Newcastle were taxed (a reflection of wealth) more heavily in 1334 (Hoskins, 1959, 174, 176; Saul, 1979, 105). Hudson (1907,183) attributed much of Great Yarmouth's wealth to the influence of the imports and exports through the harbour. It was a wool port of some significance (Saul, 1979, 112-113). In 1319, the town attempted to persuade the king to appoint it as a staple for wool exports and in the latter half of the 14th century, Yarmouth rivalled Norwich for the status, with the role fluctuating between the two towns in the last decade of the century (Saul, 1979, 113). The town also had approximately 65 ships of 100 tons (tonnes) engaged in the Bordeaux wine trade early in the 14th century (Saul, 1979, 105, 108).

Medieval Great Yarmouth exported corn and herrings and, somewhat suprisingly, ale. The town received coal from north-east England, probably Northumberland, in the 13th and 14th centuries (Tingey, 1913, 141, 143 and 147). It also imported wine; and both normal salt and bay-salt (obtained in large crystals by slow evaporation; originally, from sea-water by the sun's heat) (Rutledge, 1994). In the mid 14th century, Great Yarmouth's autumn herring fair attracted merchants from places as widespread as Spain and Scandinavia (Saul, 1979, 105). Swinden (1772, 94) records that in the five-day period of 28th September to 3rd October 1342, of 60 foreign fishing boats that entered the haven at Yarmouth, ten were from Lombardy in Italy. The importance of the herring catch to the prosperity of Great Yarmouth was such that herrings figure prominently in the municipal coats of arms of the town (Figure 1.6). Building stone was a further material both imported and trans-shipped through the port. In the latter category, Ayers (1990, 223) states, for instance, that stone was required 'on a colossal scale' for the construction and subsequent repair of Norwich Cathedral.

Great Yarmouth was also a key maritime base and defensive post. It was, in medieval times, 'one of the principal Seaports of the Kingdom, a frontier town ... the key to Norfolk and Suffolk' (O'Neil and Stephens, 1942, 3). Many townsmen were ship-owners, and in the first half of the 14th century Yarmouth was the largest provider of naval vessels north of the River Thames during the wars with Scotland and France (for further details, see Saul, 1975; 1979, 109).

Great Yarmouth's prosperity depended upon its harbour, its haven and piers. A dramatic economic deterioration began in the late 1330s, and was especially marked after the mid 1360s, brought about by a combination of natural disasters, not least the blocking of the harbour exit in 1336, and of its artificial replacement, the First Haven, about 1346 or 1347 (Gruenfelder, 1998, 143). The arrival of the Black Death in or about 1349 (Carter, 1980a, 302; Fakes, 2000, 25), the Hundred Years War with France, and a decline in the herring, wool and wine trades (Saul, 1979, 106; Rutledge,1999, 30) also helped to produce economic decline. The precise date of the arrival of the plague in Great Yarmouth is uncertain as are the numbers afflicted or how many were to die. Manship (1619, 35, 100, 180) and Harvey (1969, 183) put the death toll at 7,000. A further plague ravaged the town in 1579 (Manship, 1619, 137; Palmer, 1864a, 120), when as many as 43 persons died in one day (Manship, 1619, 137; Swinden, 1772, 100). By 1377, Great Yarmouth ranked eighteenth amongst the provincial towns in order of tax returns, having been outstripped by Norwich (4th) and King's Lynn (7th) (Hoskins, 1959, 174, 176). Hoskins made the point that his comparative figures for the two years 1334 and 1377 were obtained by different means, and they both must in part reflect the amount of merchandise in transit through the port which might have

The medieval town wall of Great Yarmouth, Norfolk, U.K.

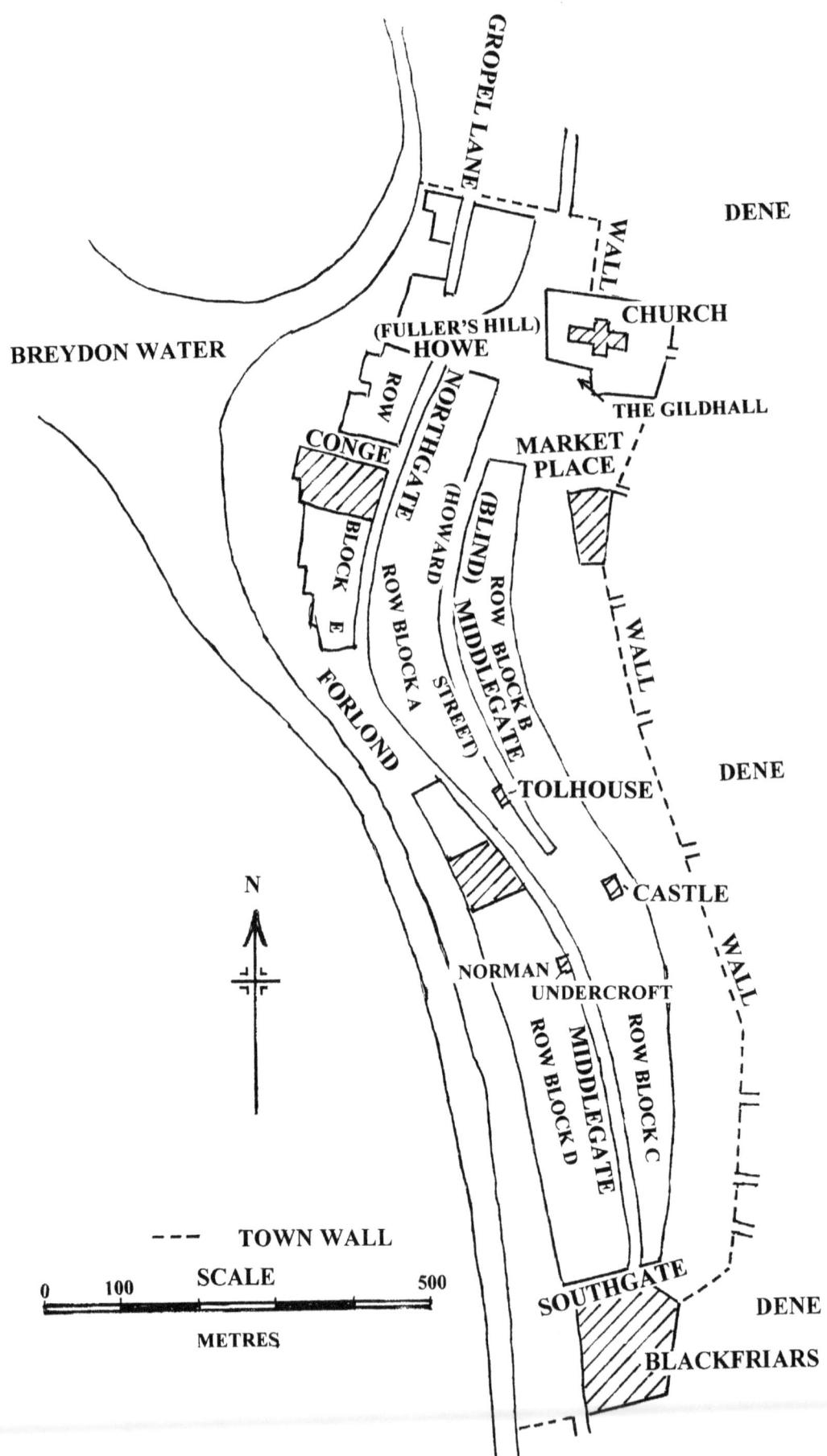

Figure 1.5 A plan of Great Yarmouth in the late 13th century, copied by kind permission from Rutledge (1990).

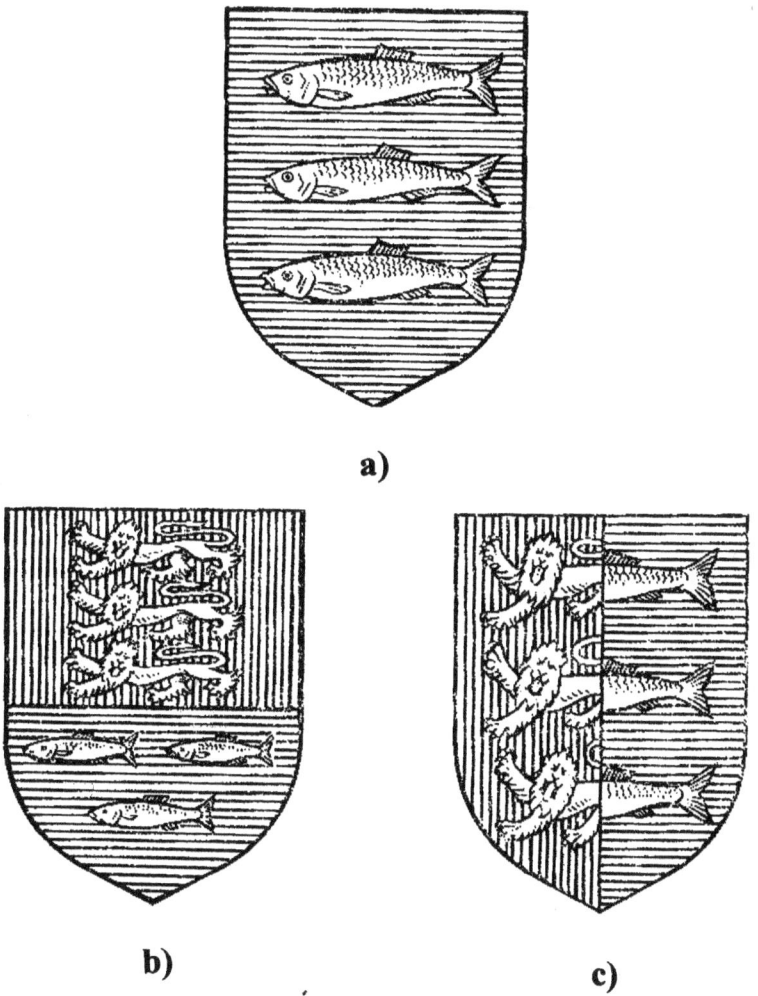

Figure 1.6 The importance of the herring industry to Great Yarmouth's economy is reflected in the town's coat of arms, which has been modified over time. The present coat of arms, adopted in 1563, is illustrated in c). Reproduced from Ecclestone and Ecclestone (1959, 101).

been abnormal for a particular year. Despite such factors the change in status was dramatic.

For most towns, this change in status was relatively shortlived; Great Yarmouth, however, on the interpretation of building and deed roll evidence (Rutledge, 1999, 30), did not enjoy a recovery until late in the 16th century. Gruenfelder (1998, 143) records that between 1549 and 1613 the town spent a total of £38,652 (at 2006 prices, very approximately £5 million) on its haven and piers. Construction of the seventh, surviving, haven between about 1560 and its final completion date of 1613, contributed significantly to the revival in the town's fortunes (Carter, 1980a, 302). For further discussion of the causes of town's decline, see Saul (1979).

1.3 Published and manuscript histories of Great Yarmouth

'Great Yarmouth is indeed fortunate in having preserved the complete records of its Assemblies from 1550 onwards.

There is also a most valuable collection of court rolls extending over a very long period.'
Ecclestone and Ecclestone (1959, 44)

A wide range of medieval, early documents and maps relating to the town and its history are known to exist. These include a variety of state documents, Great Yarmouth's Assembly Books and 14th century building accounts. Although all are in public repositories, some are moderately difficult to access. Rutledge (1976) has provided a useful guide to many of these documents; he notes for instance (p. 7), that the borough court rolls were discovered behind a wainscot of the guildhall when it was demolished in 1850. From the manuscripts and documents, a number of authors have established what appears to be a reasonably accurate historical account of the building, alteration and maintenance of the town wall. Rutledge (1976, 7) has indicated that the borough records were used as source material in the long sequence of local histories of the town, including those by three important authors referred to extensively in this work, namely; Manship (1619), Swinden (1772) and Palmer (1854; 1856; 1864a; 1872-1875). More recently, Ecclestone and

Ecclestone (1959) provided a very readable account of the town's history gleaned from these records.

Certain early documents that make reference to Great Yarmouth should be recorded. The earliest surviving history of the town was thought to be an anonymous work entitled *'Great Yermouthe. A booke of the foundacion and antiquitye of the saide towne, and of diverse specialle matters concerning the same'* which was published from an original, and temporarily mislaid, manuscript by Palmer (1847). The original document is thought to have been compiled between 1594 and 1599 (Rutledge, 1963, 120). Occasionally referred to as the *'Foundacion'* the diverse records from which the original may have been obtained and its now, generally accepted author (Thomas Damet), were discussed in full by Rutledge (1963; 1968). Rutledge (1963, 120) suggested that the manuscript was one of the sources for Thomas Nashe's *'Lenton Stuffe'* published early in 1599, which followed closely upon the *'Foundacion'*. Nashe's work was republished in *Harleiam Miscellany*, 6, in 1810; and again in McKerrow and Wilson (1958). Although Stoker (1990, 124) regarded the work as mediocre in its scolarship, Blomefield's monumental *'An essay towards a topographical history of the county of Norfolk...'* which relates to Great Yarmouth in Volume 11, provided a readable but highly derivative account of the evolution of the town's walls from their possible foundation to 1658. The work was attributed at the time of publication to Parkin (Stoker, 1990, 123-4) and reprinted as *'The history and antiquities of Great Yarmouth'* by Whittingham in 1776 (Stoker, 1988, 21).

Provided here, a moderately detailed summary as to the availability of other historic documents, how in/complete, detailed and reliable they are, where the original material is to be found, and how the sum total of documentary evidence for Great Yarmouth compares with that for other towns/medieval walls, would be in repetition of authors such as Harrod (1855), Rutledge and Rickwood (1970), Turner (1970) and Rutledge (1976). In its place, in the present work, where reference is made to historic documents, additional notes where considered relevant, are added to the bibliographic details in the list of References.

Two Elizabethan picture-maps (perspective drawings) show the complete medieval enceinte of the town prepared in anticipation of the Spanish armada invasion of 1588. These are:
1. The original of the Cottonian map (about 1585; Cott. MS, Aug. 1, 74) is now preserved in the British Library. Reproductions of the map appear in Palmer (1854, facing p. 287) and on the cover of Hedges *et al.* (2001). A further reproduction from the copy in Palmer (1854) appears in the present work as Figure 5.3.
2. The so-called Hatfield House map which includes some features which were never built and, therefore, should be used with caution. It is held by the British Museum. It is reproduced in O'Neil and Stephens (1942, plate 1) and again in Ecclestone (1971). It forms a frontispiece to the present work and is partially reproduced also in Figure 5.7. O'Neil and Stephens provide a description of the plan.

Certain other early picture-maps such as the layout of Great Yarmouth (Figure 5.8) published by Sir Robert Paston in 1668 (Rutledge and Rutledge, 1978, 112), which is held in the Norfolk and Norwich Archaeological Society Library; and Faden's Plan of Great Yarmouth 1797 (Figure 4.3), (Barringer, 1975) with permission, are also reproduced in the present publication. Both of these plans are held also by the Norfolk Record Office.

The full extent of the Great Yarmouth town wall is covered by three editions of 1:2500 scale Ordnance Survey plans:

1. First edition: Norfolk LXVI.15 (1883-1884); Norfolk LXXVIII.3 and 4/Suffolk II.3 (1884);
2. Second edition: Norfolk LXVI.15 (1904); Norfolk LXXVIII.3/Suffolk II.3 (1904);
3. 'Edition of 1928': Norfolk LXVI.15 (1926); Norfolk LXXVIII.3/Suffolk II.3 (1926).

Dates in parentheses against the first edition plans are those of the year(s) of survey; dates associated with the second and 1928 edition are those at which the original survey was revised.

1.4 The building of the Great Yarmouth town wall

1.4.1 Introduction

Great Yarmouth's government was faced with two expensive construction challenges during the 13th and 14th centuries: the provision of stable harbour facilities (for a history of the harbour and havens, see Carter,1980a, 302; Gruenfelder,1998); and the building of a military defensive system against attack by maritime enemies, bolstering the role already provided by the castle. The castle, built sometime before 1208 (Rutledge, 1990, 43), may or may not have possessed already some form of walled protection (see section **5.1**). King John granted Great Yarmouth its first charter in 1208 (Manship, 1619, 2). Carter (1980a, 300), indicated that the date was 1209. The castle appears to have become less important following the charter and it lost its military function when the walled town enclosure, which is the subject of this work, was completed. The castle was located (Figure 1.5) between rows 99 and 101 (Rutledge, 1990, 43), nearly opposite New Gate and close to where the church of St George (now a theatre) stands (Ecclestone and Ecclestone, 1959, 84). It was pulled down in 1621. In 1261, Henry III granted by letters patent licence to enclose the town with 'a wall and ditch' (*Calendar of Patent Rolls*, 1258-1266, 177). Apart from providing defence, the wall would also function as a customs barrier, preventing traders from escaping the customary tolls (Tingey, 1913, 132). Furthermore it would keep out any nocturnal undesirables.

1.4.2 The raising of funds

Murage grants were the principal source of the funds required for Great Yarmouth town wall construction (Palmer, 1864a, 107; Ecclestone and Ecclestone, 1959, 84). However, not inconsiderable sums were provided by voluntary contributions, occasional bequests and legacies (see, for example, Swinden, 1772, 76-78; Palmer,1854, 275). Chapter 6 describes in some detail the records of how and when the murage grants were obtained. There is no documentary evidence to support the tradition that the North Gate, ' the most considerable edifice of the kind in the town' (Palmer 1864a, 122), may have been or was erected at the expense of those who enriched themselves at the time of the Black Death in 1349 by burying the dead (*contra* Swinden, 1772, 85). In 1369 (coinciding with a threatened attack on England by Charles V of France) those who lived, traded or held property in the town contributed to the cost of repairing stretches of wall (*Calendar of Patent Rolls*, 1367-1370, 256; see also Tingey, 1913, 134), and this was repeated in 1385-1386, when the French again threatened England (*Calendar of Patent Rolls*, 1381-1385, 540-541, 545; 1385-1389, 135, 177, 258 and 259). At the time of the Spanish Armada (about 1588), Elizabeth I compelled the people of Norfolk, Suffolk and the city of Norwich to help defray the cost of wall repairs (Ecclestone and Ecclestone, 1959, 86). A large sum of money spent on ordnance for the fortifications in 1642 in connection with the Civil War was raised by the provision of a rate on the inhabitants (Palmer, 1854, 277).

1.4.3 The construction timetable

Work on the building of a wall appears to have been underway by 1285, since ' the burgesses ... have shown to the king that they received this sum [40 marks] and expended it and more about the enclosure of the town and the making of the ditches about the same' (*Calendar Close Rolls*, 1279-1288, 328). However, Palmer (1954, 330) indicated that no work had commenced on the wall in 1287. Furthermore, a sustained construction programme is unlikely to have started before 1321 when a lengthy series of murage grants came into effect (Turner, 1970, 141). The town's dramatic economic decline, which began in the late 1330s, is thought to have been responsible for a slowing down of work on the wall (Carter, 1980a, 302). How much of the wall had been completed by 1336 is not known (Carter, 1980b, 303); but between 1336 and 1345 around 600m. (113.5 rods, plus an additional unspecified distance in 1344-1345) were built from the foundations (Swinden, 1772, 81, 85 and 89-92). This distance, Carter (1980b, 303) believed constituted approximately one-third of the entire circuit.

A number of authors have indicated that wall building commenced at King Henry's Tower (Swinden, 1772, 83; Ecclestone and Ecclestone, 1959, 85; Carter, 1980b, 303). Turner (1970, 140) proposed instead that building took place all round the town, with the wall standing to different heights in different places throughout the first half of the 14th century. Referring to the years 1336 to 1345 when the accounts of the collectors of murage are available, Turner estimated that about 608m. of wall were built from the foundations and a further 118m. of wall were heightened. She argued that it was never precisely clear where the sections of wall being erected or raised were in relation to the total wall fabric and that 'work was in progress at both ends of the town simultaneously' (p. 141).

In 1345, labour was diverted to the construction of the First Haven (Carter, 1980b, 303). The Black Death epidemic of 1349, may (Blomefield, 1810, xi, 355) or may not (Tingey, 1913, 134) have brought about a further standstill; it was, however, indirectly responsible for a reduction in murage income and an increase in the price of labour (Swinden, 1772, 77; Tingey, 1913, 134). The town wall was stated as being greater than 2,000m. long in the 1380s (Saul, 1979, 107; this statement may be incorrect for his citations are, most unusually, here inaccurate), that is, what should have been only about 60m. short of the full circuit length, yet construction has been stated as still not at an end by 1386-1387 (Swinden, 1772, 76). Turner (1970, 140-141) provided further details on the progress of the building of the wall.

1.4.4 Wall building completion date

It is commonly suggested that work on the wall was at a stage of completion by the mid-1390s. From this, it would appear that approximately 110 years after permission to collect murage fees was first granted, Great Yarmouth's town wall had been completed. Carter (1980b, 303), for instance, proposed that the walls were possibly finished in 1393 when construction of the Second Haven began. Swinden (1772, 79) believed that with the aid of the murage grant of 1390, which was of five years' duration, the walls were finished by 1396 (see also O'Neil and Stephens, 1942, 4). He further suggested that the last part of the circuit to be finished was 'at the north end' (Swinden, 1772, 84). Palmer (1854, 275) also stated that the wall was completed by 1396, but William Worcestre, who visited the town in 1479, suggested that the circuit was finished in 1346 (Harvey, 1969, 179, 183), although he may not have meant to its full height. Turner (1970, 139), however, indicated that murage grants were 'almost continuous until 1448' and provided until as late as 1462. Some of this murage grant may, however, have been used for the acquisition of ordnance. The date of completion of the wall defences remains, therefore, inconclusive. The whole project of the erection of such substantial defences being of such a size that repair and updating work (see below) was probably concurrent with building phases taking place elsewhere in the wall.

The length of the completed structure was 2,190 yards (or 2,003m.) (Manship, 1619, 70-72), or 2,238 yards (2,046m.) (Swinden, 1772, 82, 101; Palmer, 1864a, 107; Turner, 1970, 141; Pevsner and Wilson, 1997, 489). Manship and Swinden provide details of the length of the component parts. The wall was 23 feet (or about 7.0m.) high (Manship, 1619, 73). The precise original wall length is difficult to determine today, for sections are

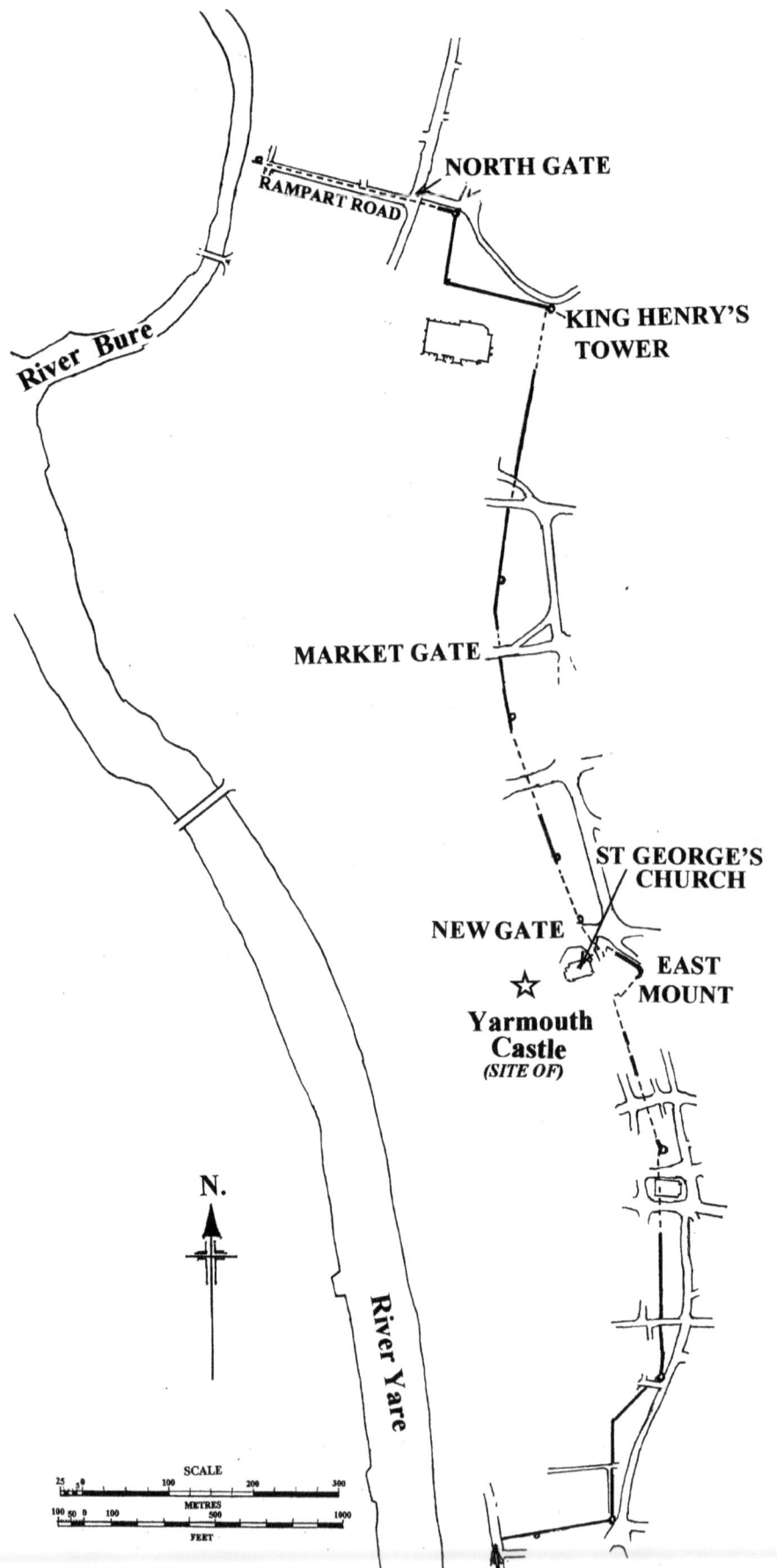

Figure 1.7 The walled town of Great Yarmouth and sites referred to in Chapter 1.

missing and the exact position of the wall unsure (as the section parallel to Rampart Road). Exclusive of the later East Mount enclosure, a figure of about 2,060m. was measured by the present author. Manship (1619, 70) refers to ten spacious gates and 16 stately towers, Swinden (1772, 96-101) to ten gates and 18 towers. Some gates were set up within towers, and these were counted as a separate towers and gate by Swinden (see also *Great Yarmouth Town Wall*, 1971, 8-9). The wall protected the town on three sides against attack from the sea, the River Yare acting as an obstacle to invasion from the landward side to the west. Together, the wall and river enclosed an area of 54 ha. (Turner, 1970, 141) (Figure 1.7).

The numerous east-facing gates, partly dictated by the shape of the peninsula and, in turn, the town, marked a 'source of weakness to the defence' (O'Neil and Stephens, 1942, 2). They were necessary to allow access to the wells, windmills, net-drying grounds and common pasture that were established on the Denes by the late thirteenth century (Rutledge, 1990, 43). The term Denes refers to the open areas of sand along the seafront, especially at the northern and southern end of the town (Sandred *et al.*, 1996, 30). The gates were defended in Tudor times, save for the North, South and Market gates (Palmer, 1864a, 108), by rampires, and structures resembling ravelins or, mounts, and artillery. The designation of gates and towers is that devised by Swinden (1772, 96-101).

> '*These walls, so high, do foe defy;*
> *Whilst Gates, so broad, do maintain trade.*'
> Manship (1619, 72)

1.4.5 Repairs to the town wall

Due to slow progress on the wall, some stretches were showing signs of decay before the entire circuit was finished. Substantial repairs were carried out in 1369 and again in 1385-1386 (section *1.4.2*).

As French naval strength grew during the course of the Hundred Years War, Great Yarmouth was ordered to strengthen its wall rather than prepare vessels (Saul, 1979, 107). Some of the most substantial modifications took place after 1512 when remission of tax was granted to allow the purchase of cannon (Carter, 1980b, 303). The towers and gates were modified to take artillery; and many of the parapets were raised to accommodate gun ports (Pevsner and Wilson, 1997, 516).

Rampires were constructed for the installation of artillery and to strengthen the wall against any possible enemy gunfire. Rampiring involved embanking the inner face of the wall with unconsolidated materials to strengthen it against artillery fire; artillery was also placed on top of the rampire. Rampiring began in 1545 on the orders of the Duke of Norfolk; the rampires were enhanced in several episodes between 1551 and 1558 (Ecclestone and Ecclestone, 1959, 49, 51, 54, citing *Great Yarmouth Assembly Book*, 1551, 1553, 1558; or from 1554 or 1555 to 1557 according to O'Neil and Stephens, 1942, 4-5, citing Manship, 1619, 73). The gates were rampired in 1553 (Ecclestone and Ecclestone, 1959, 52, citing *Great Yarmouth Assembly Book*, 1553), with the exception of the North, South and Market Gates (Palmer, 1864a, 108). The rampires in Great Yarmouth were completed to the top of the wall in 1587, and were described as being about 40ft. (12m.) in breadth from the wall (Swinden, 1772, 95-96). Internal access to the arrow-slits thus became obscured. The last episode of rampire construction is stated as taking place in 1587-1588 (Manship, 1619, 73–74; Swinden, 1772, 91-96; Palmer, 1864a, 107-108).

A number of earthen mounts or gun-platforms were also raised in the late sixteenth century (Manship, 1619, 46-48; Swinden, 1772, 96-100; Palmer, 1864a, 107-108; Ecclestone and Ecclestone, 1959, 86; Carter, 1980b, 303-304). In 1588, the East Mount, which Carter (1980b, 303) believed was erected in 1569-1570, was modified (see stretch '*G*': The East Mount, section *3.2.7*). In the same year, Sir Thomas Leyton ordered three ravelins to be constructed, but 'None...was actually constructed' (Carter, 1980b, 304). Blomefield (1810, 357), however, indicated that 'a ravelin was formed' to the east of Blackfriars as a result of this directive, of which, 'at present no vestiges remain'.

In 1625, Charles I ordered an inspection of the town's defences which resulted in the provision of a further 30 cannon. When civil war broke out in 1642, Great Yarmouth proclaimed for Parliament; the fortifications were strengthened and a new moat or ditch constructed outside the north wall (see Figure 5.9 herein; and Palmer, 1864a, 110, plan by A. W. Morant facing p. 106). Fortunately, Major General Ireton's proposal to demolish the entire medieval masonry walls in 1648 (*Norfolk Record Office*, Y/C19/7, fo. 129r – this manuscript indicates that the rank at the time was Colonel) was rejected (Carter, 1980b, 304).

Once the Civil War was over in 1651, the defences were neglected. Those remaining of the original ten gates were pulled down between 1776 and 1837 (Palmer 1864a, 118-120; Pevsner and Wison, 1997, 494). A substantial proportion of the town was destroyed by bombing in 1942-1943 (O'Neil, 1953; Carter 1980a, 302); but no part of the medieval wall was severely damaged. Approaching two-thirds (about 1155m.) of the original length of the wall is still standing, with sections up to about 7.6m. high. Eleven towers survive, eight (North West, North East, Hospital, Guard, Pinnacle, South East, Blackfriars', Palmer's) largely intact.

Table 1.1 represents an attempt to reconstruct the chequered historical account of the town wall, towers and gates.

Table 1.1 A list of the dates and events which are believed to have affected the Great Yarmouth walls, their construction and alteration.

Date	Event	References
28 September 1261	Henry III granted permission to enclose the town with 'a wall and ditch.' First murage grant (for six years)	*CPR* (1258-1266, 177)
1285	Wall building appears to be underway. Second murage grant	*CCR* (1279-1288, 328); Swinden (1772, 76). Turner (1970, 139)
?1285	King Henry's Tower constructed	Green (1970, 111) – but no evidence
1287	The wall enclosing St Nicholas's precincts is destroyed by a severe storm. 'this is before town walls were built'	Cox (1906a, 436), see also Luard (1869, 313). Palmer (1854, 330)
1290	Reference to existence of the wall at the Dominican friary site	Page (1906, 436); Carter (1980a, 302)
*c*1321	Sustained building programme begins	Turner (1970, 141)
1336 and 1337	An old wall outside the town was pulled down and the material used 'for the augmentation and expedition' of the town wall	Swinden (1772, 80, 84)
1337-1338	At work at the Black Friars and south end, and adjoining wall under construction	Swinden (1772, 83-84); Turner (1970, 140)
1341	'at work near Blackfriars'	Ecclestone and Ecclestone (1959, 85)
1342	Blackfriars' Tower completed	Swinden (1772, 89); Palmer (1854, 417)
1342-1343	Section between the South East and Blackfriars' towers increased in height	Swinden (1772, 89); Turner (1970, 140)
1343-1344	Work being carried out at the northern end of the town	Swinden (1772, 90); Turner (1970, 140)
*c*1343-1344	North West Tower erected	Swinden (1772, 90)
1344-1345	Foundations of the South Gate raised	Swinden (1772, 92)
?1348	Bubonic plague reaches Great Yarmouth	Manship (1619, 35, 100, 180); Carter (1980a, 302)
1349	Bubonic plague reaches Great Yarmouth	Swinden (1772, 77); Tingey (1913, 134); Fakes (2000, 25 [January, 1349])
?1349	North Gate erected?	Swinden (1772, 85); Palmer (1864a, 122)
1369	Substantial wall repairs necessary – townspeople contribute	*CPR* (1367-1370, 256); Tingey (1913, 134)
1385-1386	Further substantial repairs needed	*CPR* (1381-1385, 540-541, 545); *CPR* (1385-1389, 135, 177, 258, 259)
?*c*1393	Entire circuit completed	Carter (1980b, 303)
?1400	Entire circuit completed	Swinden (1772, 76); O'Neil and Stephens (1942, 4); Turner (1970, 139)
1525	Blackfriars' church (and monastery?) ravaged by fire and never restored	Manship (1619, 38); Palmer (1864a, 113); Cox (1906a, 436); Rye (1973, 498)
1535-1536	Blackfriars' monastery/friary dissolved.	Manship (1619, 38).
1539	Dissolution of Blackfriars	Rye (1973, 498)

Date	Event	References
1542	Cap added to the Pinnacle Tower	*Norfolk Record Office* Y/C18/6, fo. 29v
1542	Oxney's Tower given new roof	Rutledge and Richwood (1970, 49)
1544	Rampiring of walls first took place	Swinden (1772, 92); Carter (1980b, 303)
1544-1587	Walls rampired, especially between Market Gate and Blackfriars	Manship (1619, 73); O'Neil and Stephens (1942, 5)
1545	The Duke of Norfolk informed King Henry VIII that the Great Yarmouth wall was in need of repair. Sand dunes had filled in the moat and grown up against the walls and almost buried them	Palmer (1854, 417); Ecclestone and Ecclestone (1959, 85)
1545	South Mount first documented	*Great Yarmouth Town Wall* (1971, 4); Carter (1980b, 303-304)
1551	Repairs to the southern end of the wall	Palmer (1864a, 108); Ecclestone and Ecclestone (1959, 47-48, citing *Norfolk Record Office* Y/C5/4)
1553	Rampiring of the southern end of the wall and 'to rampire the gates'	Ecclestone and Ecclestone (1959, 51-52)
1557	Walls further rampired	Manship (1619, 73); Swinden (1772, 95)
1557	Reconstruction of wall between Friars Lane and the South East Tower, it having collapsed or been swept away by 'a great rage'	Palmer (1852, 390); Palmer (1854, 417); Rye (1973, 502)
1566	Blackfriars' Tower was ordered to be repaired	Palmer (1854, 417)
1569-1570	The New Mount constructed. The East Mount was built	Manship (1619, 46); Swinden (1772, 97-98); Palmer (1864a, 107-108); Carter (1980b, 303). *Great Yarmouth Town Wall* (1971, 4)
Pre c1585	?Flint and brick chequerwork, and conical caps possibly added to some towers	Evidence from Cottonian map of c1585
1587	Gates arched over with brick	Manship (1619, 73-74)
1587-1588	Rampiring from Blackfriars to the Market Gate 'formally finished to the top'	Manship (1619, 73-74); Swinden (1772, 93); Palmer (1864a, 108)
1588	The triangular East Mount ravelin was added to the outer face of the wall	Manship (1619, 74); *Great Yarmouth Town Wall* (1971, 4); Rose (1991)
1588	The South Gate ravelin was thrown up	*Great Yarmouth Town Wall* (1971, 4)
1588	Charnel house (at St Nicholas) used for stone source	Manship (1619, 40)
1588	North side of churchyard St Nicholas 'in part dilapidated'	O'Neil and Stephens (1942, 6, plate 1)
c1588 1590	South Mount remodelled constructed [or reconstructed]	Carter (1980b, 304); Swinden (1772, 96); [reconstructed = *Great Yarmouth Town Wall* (1971, 4)]
1590	An inner wall higher than the town walls was erected	Manship (1619, 47)

Date	Event	References
1620-1621	Great Yarmouth castle pulled down in response to great demand for building material	Palmer (1864a,117); Time and Tide Museum of Great Yarmouth Life's display panel
1626	The Main guard (later the ?Battery mount) with encompassing wall, higher than the town walls, extending to the Market Gate, erected	Swinden (1772, 99)
1633	Peter Dennys granted permission to erect windmill on Palmer's Tower	Great Yarmouth Town Wall (1971, 6)
1642	St Nicholas Tower and Gate demolished	Ecclestone and Ecclestone (1959, 90)
1642	The North Gate ravelin was constructed	Great Yarmouth Town Wall (1971, 4)
1648	Colonel (Major General) Ireton recommended the demolition of the entire wall	Norfolk Record Office, Y/C19/7, fo. 129; and rejected, Carter (1980b, 304)
1766-1837	'All ten' gates were demolished	Palmer (1864a, 118, 120); Great Yarmouth Town Wall (1971, 4)
1799	Wall removed at St Nicholas to permit graveyard to be enlarged	Palmer (1864a, 121)
1807	Passageway to 'Black Friars' gardens' cut through Blackfriars' Tower	Palmer (1854, fn, 417)
1807 or 1808	North Gate's two towers demolished	Palmer (1864a, 122); Ecclestone and Ecclestone (1959, 87) [1808]
1812	South Gate demolished	Ecclestone and Ecclestone (1959, 91)
1830	Market Gate demolished	Ecclestone and Ecclestone (1959, 90)
1837	Pudding Gate removed	Ecclestone and Ecclestone (1959, 90)
1902	The wall between the North-West Tower and Northgate Street, save for one short stretch immediately to the west of the North-East Tower, was demolished	Anon. (1980, 18)

Notes:

CPR   Calendar of Patent Rolls
CCR   Calendar of Close Rolls

1.5 Summary

There is a relatively prolific quantity of historic documentation related to the development of Great Yarmouth from the time of its earliest development, on a sand bank in the broad estuary between Roman Caister and Burgh Castle, until modern times. This evidence indicates that a wall was constructed about the town for defensive purposes mainly during the period between 1340 and 1395, but possibly continuing until as late as 1462. Precisely what was present prior to the earliest of these dates remains less clear. The 14th century wall, built with towers, gates and a wall that reflected the period, with arrow slits for archers and a sentry-walk, appears to have been faced with flints. The historic records provide some evidence of the necessity for repairs and rebuilding to the wall. In particular, the changing requiremets of warfare necessitated the walls undergoing extensive alterations in the mid to late 16th century. This involved the strengthening of the whole walled enclosure with interior earthen embankments (rampires) together with some wall rebuilding.

CHAPTER 2

THE STRUCTURE AND COMPOSITION OF THE WALLS

2.1 Introduction

Since their earliest construction, the structure and composition of the Great Yarmouth town walls prove to have been modified extensively. This detailed examination indicates that various wall sections represent different periods of construction; the modifications occurring either wholly, or in part, over time. All previous accounts as to the construction materials used in the walls appear to have been generalised to simple statements. Palmer (1864a, 106) described the wall in a manner typical of many others; as 'rubble, composed principally of Norfolk flints, interspersed with hard flat bricks, firmly united by concrete, and faced externally with smoothly-cut flints'. The details of the provenances of the construction materials are equally misleading. The wall is described by Ecclestone and Ecclestone (1959, 85), as being of 'local pebbles and flints'.

As both the structure and the composition of the walls vary throughout their thickness, in the present descriptions, in each section the interior and the exterior of the walls, as well as the composition of the wall's rubble core where visible, are described and analysed separately.

2.2 Possible construction of the walls – earlier proposals

Few attempts to analyse the structure of the Great Yarmouth walls have been made by the many authors who have written historical accounts of these remarkable defensive features. Indeed, all authors appear to have assumed that the walls as they currently stand are the remains of the original 14th century construction. Tourists examining the walls and towers today are advised by wall plaques and notices that the features which attract their admiration were built in this period. The evidence for 14th century workmanship, and how much work of this date remains, will be discussed by the present author in later chapters.

Following a small archaeological excavation (on the south side of the Market Gates shopping area, Figure 2.1) in the summer of 1955, Green (1970, 114) proposed that the town walls were enclosed in a moat. A number of early documents (as *Calendar Close Rolls*, 1279-88, 328) had certainly referred previously to work on 'ditches'. The available evidence for the possible existence of this moat and, if present, the possible date of its construction are provided in section **5.5**.

The excavations made by Green in 1955 led him to propose a possible technique for building the original walls (Green, 1970, 116). He wrote; 'A wall of this thickness (0.69m.), faced with knapped flint resting on a flagstone base, was built as a revetment to the inner face of the moat and taken to above the then surface-level. Inside this structure, a trench some 5.5 ft. (1.67m.) wide was dug parallel and in this were placed consolidated heaps of flint nodules at measured intervals to form the bases of the brick piers. On their levelled tops were set the wider brick footings and over these grew the inner brick part of the wall, the outer flintwork doubtless being continued simultaneously. In the meantime the vacant parts of this foundation trench had been refilled with some of the removed sand with which had become intermingled some of the mortar rubble from the building activity.' Green (p. 116) also described the upper portion of the wall core as being of 'grouted rounded flint beach-cobbles'. The total height of the constructed wall as measured by Green (p. 114) was 31ft. 9in. (9.68m.) of which about 12ft. (3.66m.) was below current ground level (p. 112). Green's vertical cross-section of this wall is illustrated here as Figure 2.2. Turner (1970, 139) makes the valid observation that any excess soil from a foundation trench must have been carted away for there is no sign that the original walls stood on a bank.

The wall structure was also described briefly by Carter (1980b, 303), but the information provided is very similar to that offered by Green. Although the structure of the wall as detailed by Green above ground can be confirmed, his excavation remains the sole investigation below ground at a point where the wall still stands. It is, therefore, impossible to determine whether the foundation structure as described may be universally applied to the walls, or if it is only specific to the investigation site.

Green (p.113) proposed that the wall was built continuously with, and like the adjoining moat, on a flagstone floor (Figure 2.2). The situation regarding the moat and a possible flagstone floor is discussed in section *5.5.1*. It is far more probable that the wall foundations were set on a timber framework filled with sand or rubble. An excavation on the quay in the south part of the town on the line of the wall, recorded by Crowson (1997), revealed that 'the base of the wall had been built upon a wooden frame, subsequently infilled with mortared rubble and flint'. The historic manuscript records provide some support for this type of wall foundation. For example, timber and sand were recorded by Swinden (1772, 90), as being supplied for the foundations of the North West Tower.

2.3 Building materials currently evident in the walls

The commonest component materials visible in the town walls as they currently stand are relatively restricted in number. However, detailed analyses, particularly of the brick types and the manner in which the materials have been used, enable different stretches of wall to be distinguished. Two materials predominate: flints, all

The Medieval Town Wall of Great Yarmouth, Norfolk, U.K.

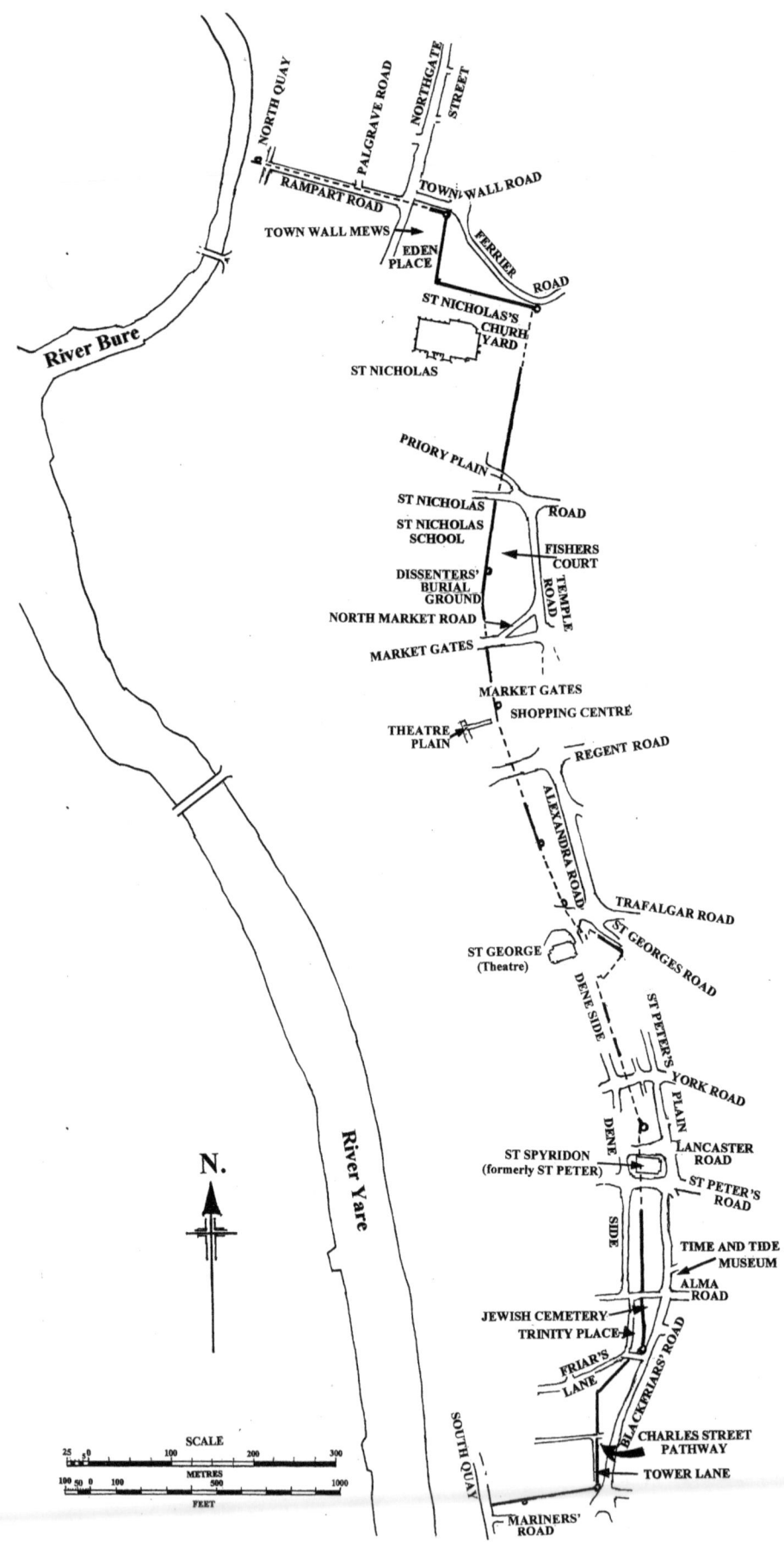

Figure 2.1 Great Yarmouth town walls and their relationships with the positions of modern localities and features referred to in this work.

CHAPTER 2 THE STRUCTURE AND COMPOSITION OF THE WALLS

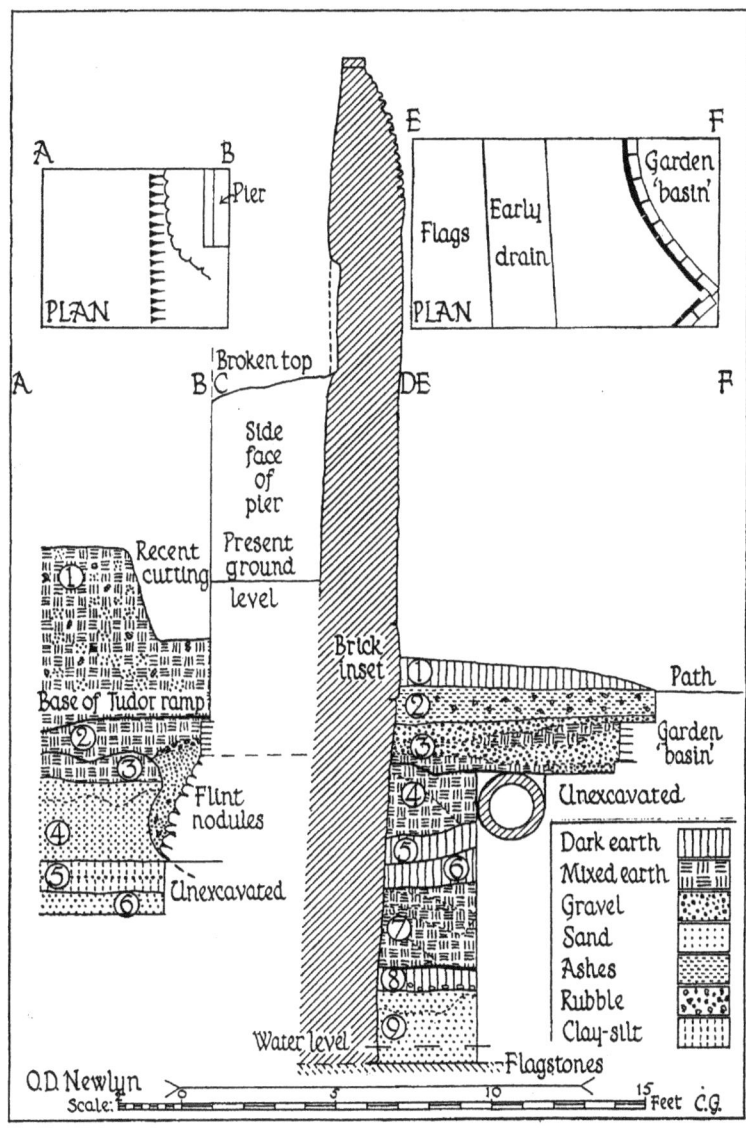

Figure 2.2 A vertical cross-section through the Great Yarmouth town wall constructed from evidence obtained from a small excavation made in 1955 on the south side of the Market Gates shopping centre. Reproduced from Green (1970, Fig. 2).

initially derived from the Upper Cretaceous Chalk (Chalk Group), and probably exclusively from the Upper Chalk, and a wide variety of whole and broken bricks. Rocks, such as flints, may be of varying sizes and in the ensuing descriptions sizes will be described in accordance with the definitions of Wentworth (1922) as; pebbles (intermediate axis length 4-64 mm.), cobbles (64-256 mm.) and boulders (>256 mm.).

Further details regarding the sources of the building materials in the walls and the some of the expenditure involved in acquiring the materials are discussed in Chapters 4 and 7.

2.3.1 Flint

Flints may be usefully described according to their provenance, their method of derivation and the manner in which they have been worked and used. Clifton-Taylor (1972), Shepherd (1972), Hart (2000) and Allen (2007) have each elaborated on aspects of the use of flints as a building stone. In accordance with where and how the flints were initially gathered they may be defined as:-

a). Mined or quarried flints: obtained directly from the Upper Cretaceous Chalk (Chalk Group). They may be identified by their nodular (but occasionally tabular or sheet-like) shapes and the presence of a white external (less silicified) cortex (Figure 2.3). Nodules are typically of cobble size. This category of flints is found in parts of the wall core and external wall facing. It should be noted that such flints require transportation from an external source, for locally the Chalk is covered by a relatively thick sequence of geologically younger deposits. Ayers (1990, 223) proposed that some quarried flints may have been trans-shipped from the Norwich area in the 13th and 14th centuries at the time of building work on Norwich Cathedral, the flints being used as ballast in otherwise empty boats. In the external facing of the wall the quarried or mined flints have generally been worked.

b). Nodular cobbles of beach flint: flints falling from a sea cliff to the beach below, or proximate to the sea as those in or adjoining a Chalk wave-cut platform, have generally suffered a level of erosion from wave action. The softer cortex may have been wholly or partially removed by erosion and the cobbles may show limited rounding. Such flints are extremely common in the wall. Their presence indicates collection from this type of sea-shore locality. A shoreline nearest to Great Yarmouth exhibiting appropriate Chalk with flint properties is that in the vicinity of, and to the north of, Cromer. Beach cobbles are easier to gather than are flints which have to be mined or quarried and typically they are used in the earlier episodes of building (Pearson and Potter, 2002; Potter, 2005a). It should be noted that because of their hardness and, in particular, their hardness relative to various forms of marine life, beach flints rarely preserve the evidence of marine life borings or attachments that may be observed in cobbles of softer rocks.

c). Rounded cobbles and pebbles of beach flint: the actions of beach drift and wave erosion gradually decrease the size and round the flint nodules described above. Long-eroded and well-rounded beach pebbles, although present in relatively small numbers on local Great Yarmouth beaches, are very uncommon in the walls. They are occasionally used for recently applied surface patching.

d). Field picked flints: Typically well-worn and stained brown or yellow from hydrated iron oxides, field picked flints are secondarily derived and found in Superficial deposits such as unconsolidated fluvial, fluvioglacial and glacial deposits. Normally they have been collected from the surface of such deposits. In the Great Yarmouth walls they are rare.

The methods of the subsequent working of flints prior to use in a wall may also be distinguished:-

i). Flints not worked: each of the four varieties of flint, a) to d) above, can be used in an unaltered state, although in the larger cobble walls a variable proportion of the flints may be broken (either accidentally or on purpose). Walls built with flints that have not been worked are occasionally patterned (for instance, elongated flints can be set in a herringbone pattern), but usually the flints in the wall are disposed irregularly, and the wall is described as being built of rubble.

ii). Broken flints: increasing proportions of the flints can be broken until, in the ultimate stage all are broken and the broken face is used to provide a facing for the wall. It should perhaps be noted that a number of earlier authors who described the wall (as Palmer, 1864a) referred to the wall as being faced with 'smoothly-cut flints'. The flints have in no instances been 'cut', rather they are always broken.

iii). Trimmed flints: the flint faces can be trimmed with increasing care to enable the flints to fit more closely in the wall. Such flints are generally described as having been knapped. The process of knapping or trimming flints is one requiring skill and can more easily be applied to flints which have been quarried or mined (*i.e.* a) above). A broken flint face may be produced by varied means (Clarke, 1935; Oakley, 1972; Shepherd, 1972). When percussion (with a hammer or chisel) or pressure is applied to a flint nodule the resulting fracture is normally conchoidal (Figure 2.4). In certain flints, the process of trimming can give rise to 'nipple-like' structures on the flint surface. These 'nipples', on the broken surfaces of flints are common on certain stretches of the external facework of the Great Yarmouth walls They may also be represented, less commonly, by reverse image 'cups' on some broken surfaces (Figure 2.5). A surface produced by several intersecting conchoidal fractures may show certain protuberances. It is suggested here that in order to diminish the size of such protuberances, the flint knapper used an instrument like a fine pointed and hard, iron chisel to remove them, resulting in a small, near perfect, cone of percussion or nipple. This interpretation differs from that of Rose (1860). Rose also first observed these structures in the Great Yarmouth Walls, where he referred to them as papillae, stating that the flint broken surface had a 'mastoid appearance'. He proposed that the nipples were the result of frost action and reflected that 'the state of the face of the flints will be some criterion of the age of the building'. He, like the present author had observed that the structures were absent from walls constructed from flints prepared by Victorian or more recent knappers. However, nipples are absent from all Anglo-Saxon flint walls observed by the present author (as Potter, 2005b; 2006), and typically, from Roman flint walls. It, therefore, seems probable that the existence of nipples on broken flint surfaces reflects a period of original flint working. Unfortunately, because of the virtual indestructibility of flints they are generally reused at times of rebuilding. The number of flints carrying nipples has for this reason been calculated for various stretches of wall, the higher numbers possibly reflecting greater wall originality (or accepting the argument of Rose, 1860, a wall exhibiting a greater period of exposure).

iv). Squared flints: the ultimate stage in knapping was to create 'squared', and in the final stage, and very rarely, 'cube', shaped blocks of flint which would fit together in a wall face. In the Great Yarmouth walls there are patches of exterior facework set in squared flints. It should be noted that nipples are not particularly common on squared flint surfaces. O'Neil (1953, 148) noted that the 'Duke's Head Hotel', in Great Yarmouth, dated 1609, was faced with squared flints.

v). Ornamental flintwork: a wide variety of ornamental work is known. The earliest examples of some varieties, such as 'flushwork', which appeared early in the 14th century, can be dated. Some ornamental use of more recent date is seen in the higher levels of certain wall towers.

Figure 2.3 A typical broken flint cobble which displays a thick, less silicified, white outer cortex. To preserve this cortex the cobble was almost certainly originally quarried or mined.

Figure 2.4 Broken flints typically exhibit conchoidal fractures as displayed on the broken surface of this flint.

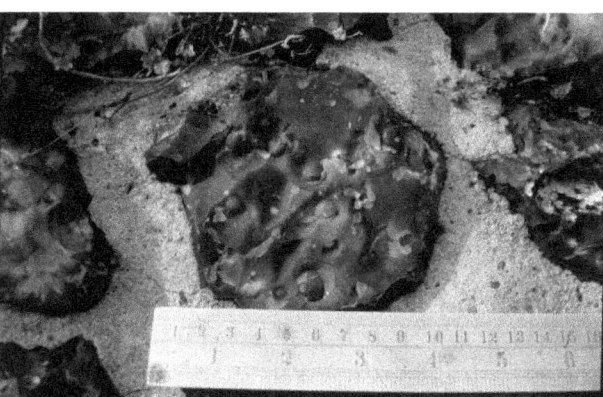

Figure 2.5 A flint cobble on which nipples are displayed on the broken surface. These are a feature of many of the broken flints in the Great Yarmouth walls.
For a discussion on their origin see section *2.3.1*.

Figure 2.6 Small conchoidal flakes of flint (created when the flint is knapped) have been inserted into the mortar between the flint cobbles in this wall face. The process in East Anglia is known as galleting, and the flakes of flint as gallets.

Further distinctions can be made in the appearance of a flint wall, the most obvious being in the quantity of mortar used in relation to flint. Anglo-Saxon flint walls of rubble, for instance, exhibit a high proportion of mortar to included flints. A technique for filling the mortar area at the wall face between the frequently irregular shapes of the flints, involves inserting the flat, conchoidal flakes of flint removed during the knapping process into the mortar. The technique, known as 'galleting' in Norfolk, was studied by Trotter (1989). In instances where the practice was thought to be datable he found the earliest to be in the period 1600-50, with the majority of the examples occurring between 1750 and 1850. Stephen Hart (pers. comm. 8th March 2007) places the earliest record of galleting in Norfolk as that at Norwich Guildhall. It, however, seems possible that this work is Victorian and was executed in 1857. Pevsner and Wison (1997, 264) date the Guildhall as 1407-1413. Hart believes the better quality examples of early galleting can generally be placed in Norfolk in the 15th and 16th centuries. O'Neil (1953, 148) regarded the galleted Yarmouth Row houses, now demolished, as dating from the 16th century. Certain stretches of the Great Yarmouth walls exhibit galleting (Figure 2.6) and they will be discussed in further detail below.

In order for the masons to create more effective and usable shapes, the present author believes that the final aspects of knapping, together with the creation of the gallet flakes, were undertaken at the wall site where the flints were used. In the instance of Great Yarmouth, the likely source of the flints (see section *4.4.2*) provides significant support for this view.

Building walls with relatively unworked flints presents, because of their shape, certain significant problems. The difficulties of constructing quoins, and of laying at the time of construction more than perhaps six courses (less in thinner walls) into freshly mixed and drying mortar, have long been recognized. The second of these complications results in certain walls exhibiting 'lifts'; recognisable levels of break for the wall to dry before building was continued. Some of the Great Yarmouth walls display these lifts (Figure 2.7).

The variations in flint walling detailed above enabled the author to develop a means (Table 2.1) by which different stretches of the Great Yarmouth walls could be differentiated. These were especially applied to the wall facings. These details were then compared, in particular, with the information provided from bricks, the other

The Medieval Town Wall of Great Yarmouth, Norfolk, U.K.

Figure 2.7 This flint wall face, which occurs in the Great Yarmouth walls in stretch 'C', displays marked horizontal breaks in the flint coursework known as building lifts. For details of their origin see section *2.3.1*.

Table 2.1 The features of flint walling used to differentiate different stretches of the flint facing to the Great Yarmouth walls.

1. Source/type of flints — mined/quarried, beach below/near Chalk cliffs, beach more distant, river/field picked
2. Style of working — whole, broken, semi-knapped, knapped, squared
3. Galleting — absent, present
4. Spacing of flints — widely spaced, moderately spaced, tightly packed, space filled with gallets
5. Quality of coursework — random (absent), poorly coursed, well coursed, (lifts visible)
6. Instrument used for working — producing nipples and cups, largish conchoidal fractures from edges of flints, no real evidence
7. Other included rocks in wall — none, ex building stones, ballast, glacial erratics
8. Other features of wall — for example, put-lock holes.

abundant building material. Together, these analyses assisted in giving suggested dates for individual stretches of wall building or rebuilding.

2.3.2 Bricks

Where the blank vaulted arched arcade remains preserved on the inside of the town walls whole bricks have been employed for the construction of its piers. Whole bricks are also present in the towers and they have been used for special features (such as quoins) on the external facing. In the wall core they are generally uncommon, but bricks which are believed to be reused often occur in vast numbers. These fragmentary bricks, frequently of about half size, tend to be used irregularly and not to provide levelling courses (as might be seen in Roman or Anglo-Saxon walls).

A variety of brick types can be distinguished in the walls. They can be identified by examining features such as the detail of their composition and method of probable manufacture, and their shape and size (Harley, 1951; 1974; Ryan, 1996; Potter, 2001). The brick colours aid in brick distinction in unusual cases only (Minter, *et al.*, in press).

The variety of brick types in the Great Yarmouth walls can be detailed as:-

TYPE 1. Late Saxon-early Norman (*c.* 11th century); these are rare, they have only been observed in a fragmentary state at Fishers Court and at the North West Tower, and they only occur in the wall core. Mean size (of eight, and less than five only for the stretcher and header measurements) 42mm. x 115mm. x 230mm. Typically, bricks of this period are about 45mm. thick. Colour: generally strong reds.
Where seen to best effect: Fishers Court, in wall core.

TYPE 2. 1300 to 1350; in places these bricks are abundant, but elsewhere they are rare or absent. They are generally broken and found in the wall core. Mean size (of in excess of 50, but of only 37 stretcher measurements) 65mm. x 134mm. x 236mm. Colour: typically red-brown through to yellow, often streaky. These bricks were probably made locally from fluvial and

or estuarine clays. They often display the impressions of straw where they were laid during the drying process (Figure 2.8).
Where seen to best effect: St Nicholas School, wall arcading piers.
TYPE 3. 1400 to 1450; these thin bricks are normally fragmentary and they vary in abundance (Figure 2.9). Mean size (of in excess of 30, but only 10 to 20 stretcher and header measurements) 43m. x 108mm. x 231mm. Colour: generally reds, more rarely through to yellows.
Where seen to best effect: Market Gates (toilet footpath) wall piers.
Very similar, but slightly thicker, fragmentary bricks may occur together with this Type 3 brick. The slightly thicker and longer bricks probably date from the earlier decades of the 15th century. The mean size (of 17, but only 5 to 10 stretcher and header measurements) being 50mm. x 107mm. x 236mm. The Type 3 bricks were also probably made from local, river or estuarine clays.
Where seen to best effect: (Thicker Type 3), North West Tower, west wall core.
TYPE 4. about 1400; yellowish, imported, possibly Flemish, bricks. All appear broken, they are never abundant and they are mainly observed in the wall core. Mean size (of 20, but only 5 to 10 stretcher and header measurements) 56mm. x 111mm. x 235mm. Bricks of this nature tend to be relatively soft, they are difficult to identify and possibly some may date from as late as the early 16th century.
Where seen to best effect: Fishers Court, in wall core.
TYPE 5. late 15th to early 16th century; Flemish or Baltic, 'clinker bricks'. Used only in a small number of re-built arcade arches. Mean size (of ten only) 50mm. x 101mm. x 202mm. Colour: yellows.
Where seen to best effect: Only seen north end Dissenter's Graveyard, in wall arcading.
TYPE 6. early 16th century; occurring mainly in arcade arches. Mean size (of 34, of more than 20 headers but only 4 stretchers) 50mm. x 114mm. x 233mm.
Where seen to best effect: Market Gate (toilet footpath), in arches around arrow slits.
TYPE 7. 1530 to 1560; these bricks are abundant particularly in walls near to Blackfriars' Tower at the south end of the town. Mean size (of 52, but only of 15 to 30 stretcher and header measurements) 55mm. x 115mm. x 231mm.
Where seen to best effect: West of Blackfriars' Tower, in remnants of wall arcading.
There are many intermediate brick types between Types 6 and 7. Colours: a wide range of reds especially.
OTHER TYPES. In particular Georgian, Victorian and more modern bricks used in the extensive repairs to the walls. These make up much of the brick work above the level of the arcade walkway.

Figure 2.8 The detail of two Type 2 brick fragments. Straw or similar material markings, created in the clay drying process, can be observed on the face of the larger fragment. The bricks are larger than those of Type 3, illustrated in Figure 2.9.

Figure 2.9 Type 3 brick fragments: these are thinner bricks (one, about 40mm. thick against the scale). The brick on the right has two unusual 'nail hole' markings (top), possibly, but with no certainty, inserted to assess the moisture content during the drying process.

Table 2.2 The information collected in the field to indicate the relative numbers of different brick types seen in sections of the Great Yarmouth walls.

Portion of wall (if visible)	Overall % bricks *cf.* other materials	Relative percentage of different brick types						
		Type 1	Type 2	Type 3	Type 4	Type 5	Type 6	Type 7
Inner face								
Core								
Outer face								

In very many instances it is not possible to provide precise dates for the bricks observed. This applies particularly to the many brick fragments which occur in the walls. The various Type 3 bricks are notably difficult to date with precision and possibly the range of dates may extend to as late as 1480. There are few known dateable buildings within this period from which comparisons of date, brick type and size can be drawn. The unusual brick with a 'double-margin' or 'sunken margin' (see section *3.2.3*) has been observed in walls of different ages elsewhere (as at Castle Acre). In Great Yarmouth walls it is only found (to date, at just one locality) only in Type 3 bricks of the shorter length, and is thought to date from 1430-1450.

Table 2.2 illustrates the document used for recording the differences in brick quantities and types which occur in the various portions of the walls.

2.3.3 Reused materials

Substantial quantities of construction material were periodically made available as existing, local buildings were demolished. Many of the bricks, particularly those used in the core of the wall, probably became accessible in this way. Type 1 bricks no doubt originated in the wall by this means. A number of structures are known, or are recorded, as having been contributory to the building materials of the original wall, its repair or reconstruction. Those worthy of record are:-

a). An earlier town fortification: Swinden (1772, 80, 84) records that in 1336 and 1337, an old wall (possibly from an earlier fortification) situated outside the town, was pulled down and the material used for the 'augmentation and expedition of the town wall'. Those who demolished the wall and transported the material were paid at least £6 2s. 11d. (approximately £3,000 in 2006 terms).

b). Blackfriars' monastery and church: A Dominican monastery was established at 'Gernemuth', on an area of land measuring 500feet (*c.* 150m.) by 500feet (equals 2.32ha.) described as 'La Straunde' (or 'la Strande', Page, 1906, 435), about 1271 (*Calendar of Patent Rolls*, 1266-1272, 536; Rutledge, 1990, 44-45). Rye (1973, 498) indicates the date might have been about 1260. A convent was added to the site by 1273 (Palmer, 1854, 409; Harvey, 1969, 185) or 1280 (Harvey, 1969, 183). Rye (1973, 498) and others (Page, 1906, 435; Harvey, 1969, 183) suggest that the church, dedicated to St Dominic, which the present author believes was imposing and large from the evidence of the masonry found in the town walls, was completed by 1280. This church, and possibly much of the monastery, was consumed by fire in 1525 and never restored (Manship, 1619, 38; Palmer, 1864a, 113). The monastery was dissolved in 1539 and the property transferred to secular hands (Rye, 1973, 498). By the early part of the 19th century the site was built over 'with streets of houses and fish-houses' (Rye, 1973, 498).

c). About 1276, the Carmelite Friars built a church possibly (Rutledge, 1990, 45) in a place called 'La Denne' in the northern part of the town (*Calendar of Patent Rolls*, 1272-1278, m. 26, 138). On 1st April 1509, this church was burnt down (Palmer, 1854, 426: Page, 1906, 437).

d). Charnel house and chantry, St Nicholas's churchyard: material from these early 14th century buildings (Manship, 1619, 40) was used in 1588 to repair the town wall, and to enclose in brick and 'free-stone' the lower part of the East Mount to form a pseudo-ravelin (Palmer, 1864a, 107-8; O'Neil and Stephens, 1942, 6). This occurred subsequent to the time of the first construction of the East Mount, when the east wall was broken through to create a canon emplacement (Manship, 1619, 46-47).

e). The castle: In 1620, the upper storey of the castle was removed and the material used for enclosing the East Mount: what remained of the castle being pulled down in the subsequent year (Palmer, 1864a, 117), in response to the great demand for building material (*Time and Tide Museum of Great Yarmouth Life*, display panel).

A number of other early buildings are known to have existed in Great Yarmouth, as for example the Greyfriars' Monastery which was known to have existed in 1271 (Coad, 1980). Although Palmer (1854, 423) states that, in 1579, 'people were permitted to remove stones for various purposes' from the church, which was probably then in ruins, there is no evidence as to what these stones were geologically, or that this material was used for repairs to the town walls.

2.3.4 Other Rock Types

A very wide range of rock types (described collectively here as 'exotics') occur in minor but variable quantities in the town walls. Only a small number of these appear to have been included, either purposefully or casually, in the original fabric of the standing walls. Most of the rock types must be considered as later additions, having been used, with but extremely rare exceptions, only in repairs to the outside and inside facings of the wall. Far fewer of these replacement stones can be observed for this reason in those wall surfaces that, until relatively recently, have been hidden by attached buildings. This difference is well exemplified when a wall such as that to the west of Blackfriars' Tower (to Palmer's Tower) is contrasted with that to the north of St Nicholas Church graveyard (to the west of King Henry's Tower). The former wall is completely enclosed in Victorian buildings on the 1884 Ordnance Survey plan and currently rarely includes rocks other than flint: the latter includes numerous other rock types and in 1884 was free from Victorian building enclosure.

Typical and more common rock types that may be viewed in the walls are listed with their locality in Table 4.3, where those identified are summarized. Probable rock types included in the original fabric of the present standing walls are:-

Chalk: blocks and small pieces of Upper Cretaceous Chalk (Chalk Group), most, if not all probably being

Upper Chalk, occur in small amounts in various parts of the visible wall core. These may well have arrived on site when occasionally included with the transported flints. Small fragments of Chalk are included also in the early lime rich mortars. Small pieces of iron furnace slag and vitreous brick slag may also be observed in certain limited areas of early wall core fabric.

Caen Stone: this cream, fine-grained, Middle Jurassic, Bathonian limestone from Normandy, France, has been described by a number of authors as having been imported to be used to dress arrow slits, loop-holes, quoins and embattlements (Palmer, 1854, 275; Turner, 1970, 141, Carter, 1980b, 303). The principal occurrence of this stone observed by the present author, where the Caen Stone was not reused, was in the earliest visible quoins of the octagonal King Henry's Tower. This distinctive tower is traditionally claimed to be the earliest structure within the wall, although there is little or no historical evidence for this tradition. The small bastion at the other end of the wall to the west of this tower also contains some early Caen Stone quoin stones.

Portland Stone: this white, Upper Jurassic oolitic limestone, sometimes of the fossiliferous 'roach' variety, is used to dress some of the gun slits which replace many of the earlier Caen Stone gun slits in the South East Tower and the adjoining wall. This tower and portion of wall contains the only obvious gun, as opposed to arrow, slits in the whole wall circuit. There is, however, no visible evidence that the Caen Stone structures may have replaced earlier arrow slits in the same positions, and this Caen Stone may also have been built into the original present fabric.

Various blocks of reused masonry are visible in the wall surfaces (especially the external facework), particularly close to the one-time site of the Blackfriars' monastery and church. These are mainly of Caen Stone, some as large quatrefoil column roundels, Barnack Stone and *Viviparus* limestone ('Purbeck Marble'), the Lower Cretaceous, gastropod rich limestone from Purbeck in Dorset. The *Viviparus* limestone blocks were almost certainly polished when present in the earlier church and various fragments of tomb slabs may be observed. The Barnack Stone is sometimes ornamented, and occasionally burnt; its blocks are of Middle Jurassic, fossiliferous, oolitic limestone. A few pieces possibly of Upper Carboniferous, micaceous sandstone in the walls may have once been floor slabs in the monastic buildings.

Glacial erratics: rock types that originated from regions to the north of Great Yarmouth, that are of broadly irregular shape and generally used for repair work to the wall's external face are probably of glacial or fluvioglacial origin. They typically include cobble-size pieces of Lower Chalk, Lower Cretaceous sandstones, and sandstones and limestones from the Jurassic. On one block only (actually on a flint with thick cortex) was there clear evidence of glacial transportation in the form of glacial striae (see Figure 3.17).

Rocks derived as ship's ballast: an enormous array of rock types, scattered in varied quantities, occur as included or replacement stones on the wall facings. Many are exotic to the United Kingdom and a considerable quantity is probably of Scandinavian origin. The rocks are typically of large cobble size, and probably originally gathered for use from beaches, for they are often water-worn to a sub-rounded or rounded shape. In order of abundance, they occur as igneous, metamorphic and consolidated sedimentary rocks. Individual rock types will be referred to in the respective wall sections as they are described.

2.4 The architecture of the wall

In order to avoid subsequent repetition, a number of points of a general nature about the architecture and wall construction are now made. The available evidence suggests that the original portions of the now-standing enclosing town wall consisted, throughout its length, of a blank vaulted arcade of Early English (Gothic) style brick arches springing from brick piers (Figure 2.10). These together supported an internal sentry-walk and protective parapet. The wall was broken at intervals only by various towers and gates. The arcading and the wall were built together and securely bonded. Each internal bay had a crossbow loophole, the outer face of which was usually dressed in (sometimes broken) brick. The arch interior lining is normally of a mixture of flints and broken bricks.

Although it may have been suggested that the sentry-walk, and therefore the arcading, was not continuous around the circuit of the walls; the present author believes that wherever the wall shows any semblance of its original form, the sentry-walk is at least in part preserved, and from this evidence is inclined to support its continuity. Manship (1619, 73) indicated that the gates were arched over in 1587, 'so that many men may walk…all alongst the walls'.

The vaulted arches (the firing bays, described by Green (1970) as 'embrasures') are approximately 2.2m. wide (Figure 2.11, A) with the piers normally between about 1.0 to 1.2m. in width (Figure 2.11, B). When constructed the arches must have been built over a timber frame. Swinden (1772, 89) described these frames as cynters. The expectation would have been for a number of timber frames to have been used repeatedly throughout the length of the entire wall (or even over stretches). Surprisingly, the present author has not found any evidence of such economy of use, for the arches seem to show unrelated variations in width. The arch bays vary in depth (Figure 2.11, C) from about 1.30 to 1.73m. This measurement approximates to the width of the sentry-walk above. The height of this walkway from the ground (Figure 2.11, D) is variable, reflecting changes in the ground level since the wall's first construction, but today rarely exceeds about 2.15m. Manship (1619, 73) gave the total original height of the wall as 23feet (7.01m.). The thickness of the outer wall varies enormously from as little as 0.65m. to more than one metre (Figure 2.11, E). The outer skin of a single layer of flints, according to the size of the flint cobbles, normally makes up at least 0.15m. of this figure (Figure 2.11, F). Green (1970, 116) gave the thickness of the wall outside the arches as

Figure 2.10 Arched firing bays observed to the north side of Garden Gate (stretch 'J'). In this stretch of wall the firing bays have been removed almost to the outer wall and, in the first two rebuilt bays, a new brick wall infills the area which would have included the arrow slit.

0.69m. and the total height of this wall (p. 114) as 31ft. 9in. (9.68m.) (See section **2.2)**. Wall rebuilding has clearly influenced all these measurements and particularly the wall thickness and height.

There are various occasions in the walls where the difference in ground level on either side of the wall is clearly indicated by the visible height of the arrow slit. Most of the arrow slits have, however, been reconstructed. In a few instances, as parts of wall '*D2*', this reconstruction is less obvious. In such examples it might be suggested that the ground beneath the firing bay was raised prior to the original construction of the wall: in which case the excavated wall foundation materials have been used for this purpose.

Most of the 18 principal towers built along the length of the wall were 'D'-shaped in plan, presenting their flat face towards the town. The North West and North East Towers were circular and King Henry's Tower octagonal. The North Gate's twin towers, which were demolished in 1807 (Palmer, 1864a, 122), were square. With the possible exception of King Henry's Tower, there is visible evidence in all the towers which remain standing, that entry could be gained to each from the sentry-walk. All traces of original stair access points to the sentry-walk appear to have been lost.

CHAPTER 2   THE STRUCTURE AND COMPOSITION OF THE WALLS

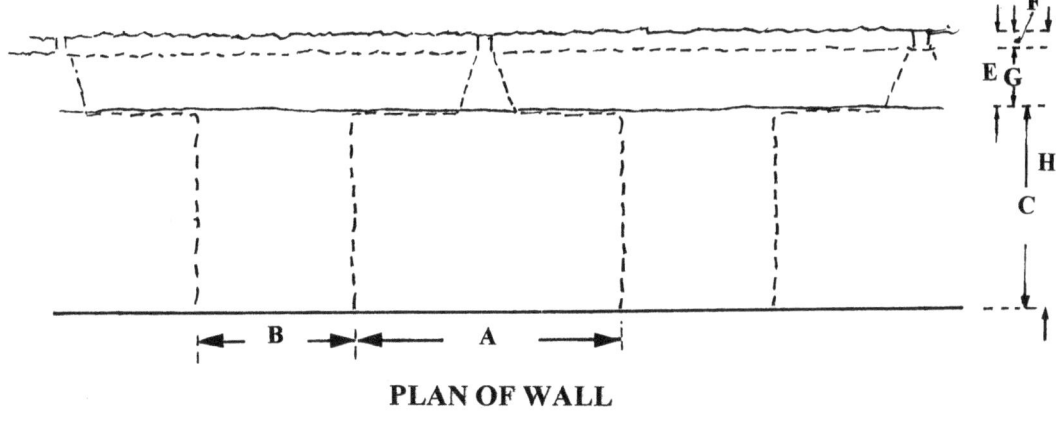

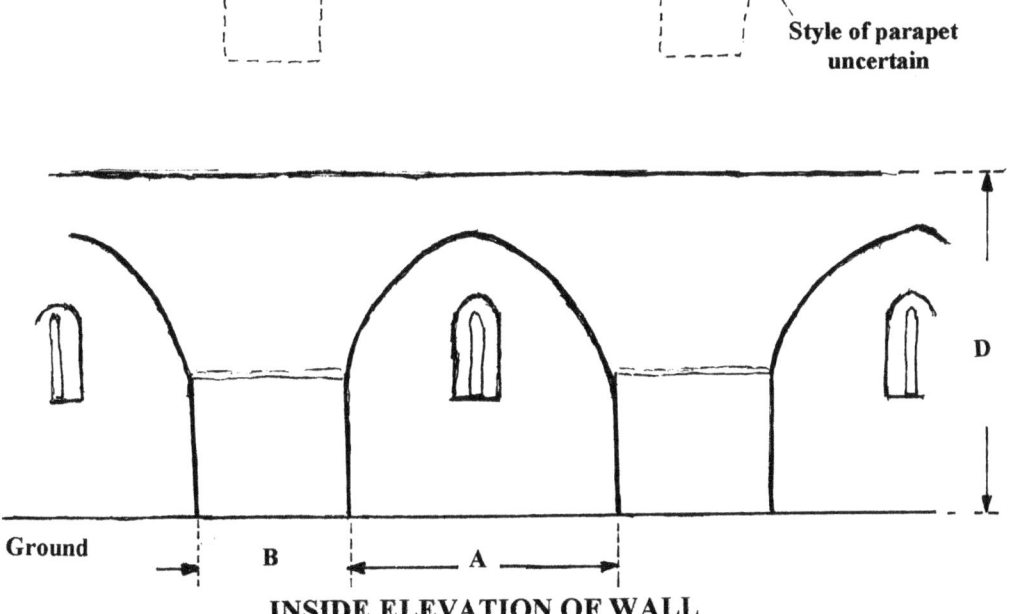

KEY
- A - Width of firing bay
- B - Width of pier
- C - Depth of bay (c. About width of sentry-walk)
- D - Height to walkway
- E - Thickness of protective wall
- F - Thickness of flint external face
- G - E less F
- H - Total wall thickness

Figure 2.11   The structure of the visible portions of Great Yarmouth town wall and key measurements, related to the wall, which have been recorded in this work.

CHAPTER 3

THE TOWN WALL: 'A MODERN GEOLOGICAL PERLUSTRATION'

3.1 Introduction

The following perambulation traces the Great Yarmouth town wall from the North West (eighteenth) Tower to the former position of the South Gate. Figure 3.1 details the names customarily used to designate the various towers and gates exhibited in the wall. The figure should be used in conjunction with Figure 2.1 to relate the positions of these features with modern roads and important town localities. The numerical designation of gates and towers devised by Swinden (1772) is not followed as many of the gates and towers have been demolished. Furthermore, at least one gate, Garden Gate, was not constructed until about 1636. Swinden described Garden Gate as his second gate. Swinden's designations are, however, referred to occasionally where they can act as useful reference points. The plan of the town which was published by O'Neil (1953) and was stated as originating from Swinden in 1738 is reproduced here as Figure 3.2. This plan is thought to more correctly be dated to about 1758 (Rutledge, pers. comm.).

Far from all the wall length remains (Table 3.1); by 1884, 710m. had been demolished. Sections that had by that time disappeared included the stretch parallel to Rampart Road in the north, the area adjoining what is now the Greek Church, erected as St Peter and constructed in 1830, and the area in the vicinity of the East Mount on which building of this mount began in 1569 (Manship, 1619, 46). By 1928, a further 145m. of the wall length had been removed. At the time of this current examination further short stretches have been lost over the last 80 years, both in the centre of the town and, for instance, to the west of the position of South Gate.

Table 3.1 A comparison of the wall length particulars between the years 1883-1884 and 1928 based on the first and third editions of the Ordnance Survey plans. The wall is taken as 2060m. in length.

Particulars	In 1883-4	In 1928
Length of wall missing (described as 'site only')	710m.	855m.
Percentage of total length	34%	42%
Length of existing wall not visible inside	760m.	782m.
Percentage of existing wall not visible inside	56%	65%
Length of existing wall not visible outside	666m.	635m
Percentage of existing wall not visible outside	49%	53%

Large amounts of the remaining lengths of wall, 1,350m. in 1884 and 1,205m. in 1928, were excluded from view by enveloping buildings (Table 3.1). Over this period of 44 years the loss of wall exposure increased in percentage terms. Largely as a result of post-World War II slum clearance, very many of the late-eighteenth and nineteenth century buildings which obscured both faces of the wall have been cleared. In Palmer's account of the defences, the tower which now bears his name (Swinden's second tower) was 'so surrounded by houses as not to be easily reached' (Palmer, 1864a, 112). The present examination reveals that there still remain many sections where access to the inner and outer faces of the wall is impossible. Many visible sections also show signs of damage where inadequate remedial work has been undertaken.

Major modifications made to the defensive structures probably in the 16th century and more recently will be discussed in Chapter 4. The names of sites that have been demolished will be referred to only where they might be useful reference points along the wall.

In this work, stretches of the wall circuit that illustrate broad similarities in structure and composition are given an alphabetical prefix ('A' to 'M'). Figure 3.3 illustrates the positions of each of these stretches and any subdivisions which they may contain. These prefixes have been introduced in order to specify localities more simply.

3.2 A description of the wall

3.2.1 Stretch 'A', North West Tower and wall fragments

The North West Tower (the eighteenth tower of Swinden) is the first tower at the north-west end of the wall (Figures 3.4 and 3.5), situated on the North Quay, beside the River Bure. Currently the wall projects towards the river 2.3m. from the point where it abuts the tower. How much further the wall extended towards the river and how it terminated is difficult to ascertain. The 1884 Ordnance Survey plan suggests that the wall once continued more than 10m. to reach the edge of the river. A 2.4m. long fragment of wall on the eastern side of the tower is all that remains of the stretch of wall that continued towards the east.

There is no evidence of vaulted firing bays; but some indication of a sentry-walk exists in the fragments of the wall that are present. The existence of a sentry-walk is revealed also in the traces of doors into the tower (the doors are about 3.5m. above ground level). The west doorway has at some fairly recent period been altered to lead down to the outside of the wall and the top two (possibly Devonian) flagstone steps remain.

Both beach flints and quarried/mined flints make up the common constituent of the walls. These are of cobble size, but in the original north external face of the wall, later inserted flints have been worked to provide improved and closer packing. Bricks are not very

CHAPTER 3 THE TOWN WALL: 'A MODERN GEOLOGICAL PERLUSTRATION'

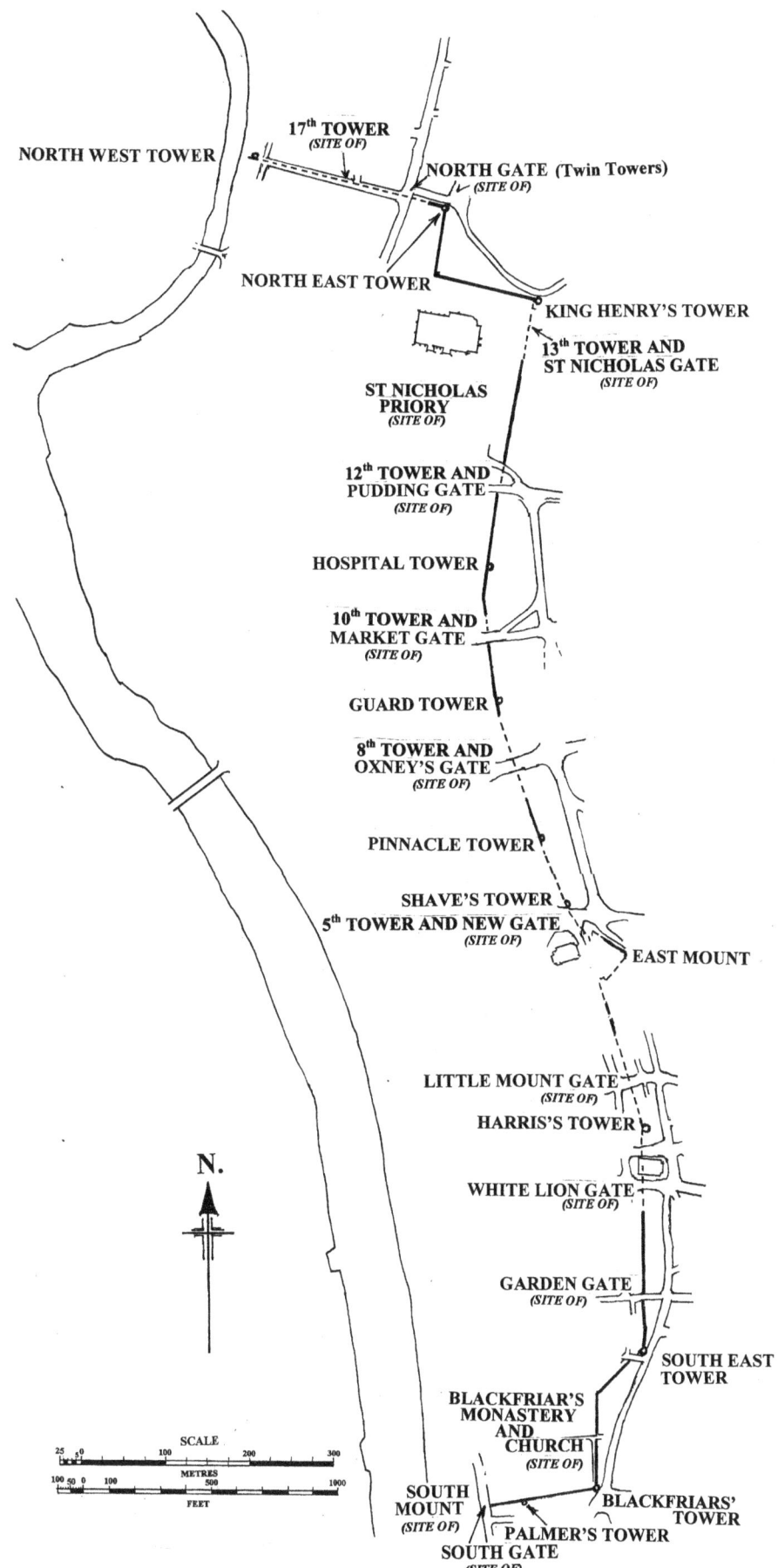

Figure 3.1 The positions and the customarily used names for the towers and gates of Great Yarmouth town wall.

The medieval town wall of Great Yarmouth, Norfolk, U.K.

Figure 3.2 Part of a plan of Great Yarmouth produced to accompany Swinden's work of 1772. North is to the left. This copy has been reproduced, with their kind permission, from The Society of Antiquaries of London (1953). The map is thought to date from about 1758, rather than 1738 as proposed by O'Neil (1953).

CHAPTER 3 THE TOWN WALL: 'A MODERN GEOLOGICAL PERLUSTRATION'

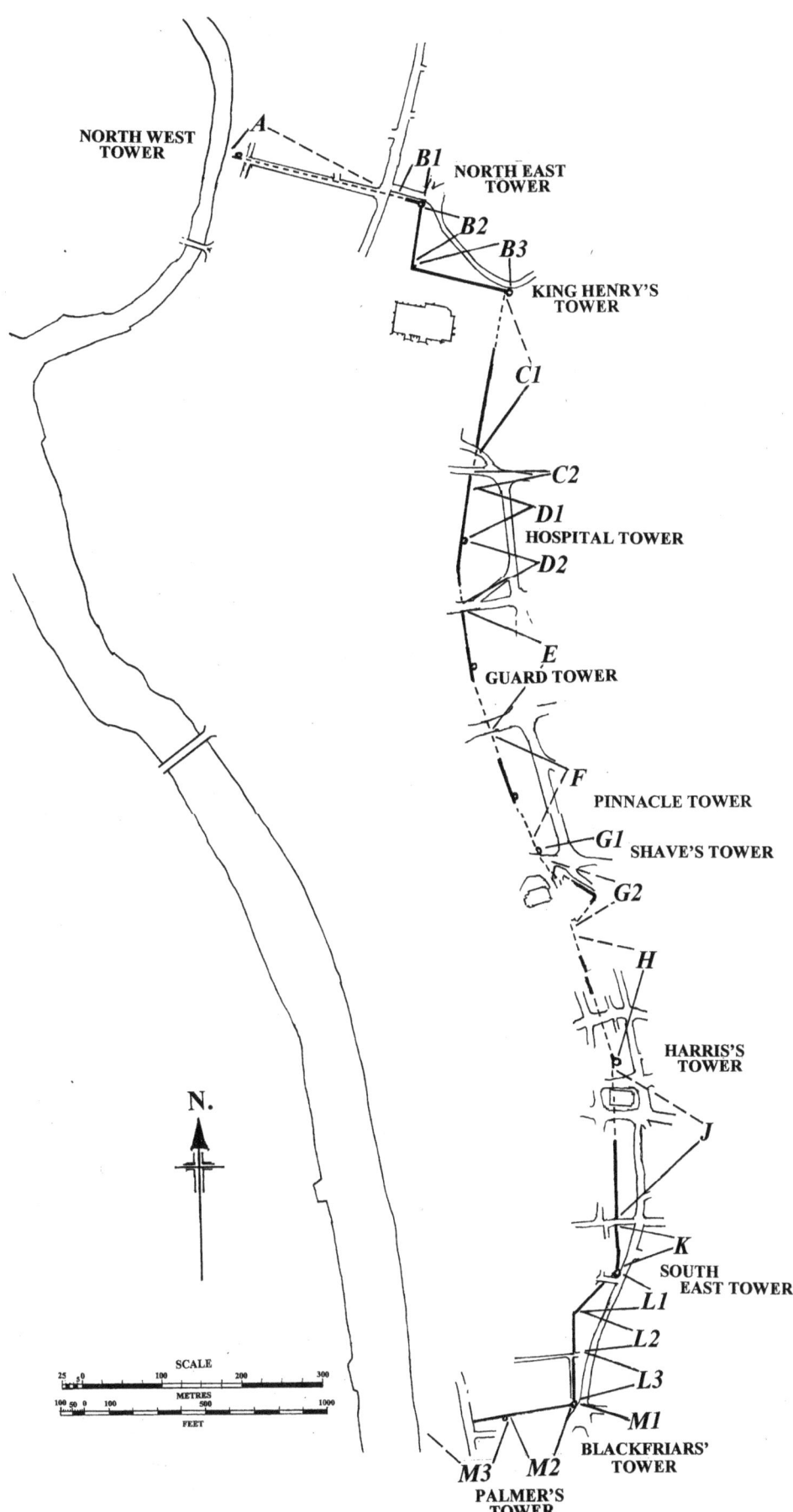

Figure 3.3 Great Yarmouth town wall and the positions of wall stretches 'A' to 'M', used in the description of the walls within the present work.

Figure 3.4 The North West Tower in about 1818. This copy of an etching by John Sell Cotman was kindly supplied by the Norwich Castle Museum and Art Gallery.

common in the wall core and are fragmentary (see Table 4.1). Although some bricks are older, the majority probably date from the mid to late 15th century. The inner wall is much repaired. It is largely built of bricks similar to the core. A single piece of Upper Jurassic grey shale is included in the broken wall core stub and is probably a recent addition at time of repair.

The 'D' shaped tower, said to be 9.14m. high, was possibly erected originally in about 1344 (Turner, 1970, 144). Swinden (1772, 90) provided information regarding the supply of materials for the tower's foundations in 1333-1334. It has been much repaired, recently in about 1960 (*Great Yarmouth Town Wall*, 1971, 16) and again about 1990. The flat face of the tower is now constructed of 19th and 20th century bricks. The third storey, 'entirely composed of thin red bricks' (Palmer, 1864a, 123) appears to represent an early 16th century replacement which has, in turn, been extensively repaired. Today, the tower is capped with a conical tile roof which has recently been repaired. The tower is in use as offices and none of its openings externally bear evidence of 14th century originality. The external flint face of the lower two floors of the tower exhibits a variety of styles of flint-work, from work similar to the immediate wall face which is consistent with an early 16th century date, to early subsequently repaired work in reasonably well-squared flint courses and recent work in patches of small flint beach pebbles. Two small fragmentary short buttresses (2m. high) project from the north side of the tower, and these are probably fairly recent additions although they do contain bricks similar to the wall core. Internally, an early arch is present inside the ground floor door and on the first floor the guardroom chimney arch has survived (*Great Yarmouth Town Wall*, 1971, 16). An early 16th century date can be applied to both, but each contains also reused bricks of the early 14th and the 15th centuries.

To the east of the North West Tower a wall along the south side of Rampart Road, although flint faced is of recent construction. Based on the position of the seventeenth tower (of Swinden), which was not demolished until 1902, the original wall was closer to the

CHAPTER 3 THE TOWN WALL: 'A MODERN GEOLOGICAL PERLUSTRATION'

Figure 3.5 The North West Tower viewed from the north. The flint facing shows considerable variation in style (see section *3.2.1*). At mid-tower height the flint courses are reasonably well squared.

Figure 3.6 The exterior of the old North Gate and the early spire of St Nicholas Church as viewed from the north-west. The gate was demolished in 1807 and this etching appeared in Preston (1819) from which this figure is reproduced.

north side of the road, for this is how its site is detailed on the 1884 Ordnance Survey plan. This tower (sometimes referred to as Ramp Tower) was situated almost directly opposite the back passage to the houses on the west side of Palgrave Road (close to the corner of the Probation Service building). The North (tenth of Swinden) Gate (demolished 1807 (Palmer, 1864a, 122) or 1808 (Ecclestone and Ecclestone, 1959, 87)) formed a passageway through the sixteenth towers (of Swinden) which were both imposing and square (Figure 3.6; O'Neil and Stephens, 1942, plate 1). Swinden (1772, 78) believed that this gate and the adjoining north wall was originally built in 1349. A plan of the North Gate was drawn by Desmaretz in 1734 (P.R.O. WO 55/1548/15, cited in Turner, 1970, 144, 146).

3.2.2 Stretch 'B', Town Wall Road, North East Tower, King Henry's Tower

For ease of description stretch 'B' has been divided into three parts:-
'B1' – a short and much reconstructed west-east orientated section that runs along the south side of Town Wall Road, together with the small North East Tower.
'B2' – the north-south aligned portion of wall which displays 20, slightly irregularly spaced, reconstructed arrow slits or gun ports. The section terminates at a small rectangular bastion (*Great Yarmouth Town Wall*, 1971, 14).
'B3' – the stretch between and including the bastion and King Henry's Tower. This follows a west-east line along the north side of the graveyard of St Nicholas Church.

Stretch 'B1', on its north external face, adjoining Town Wall Road commonly includes cobbles of ballast. Rock types such as granite, gabbro and basalt are present. Other rocks such as Jurassic and/or Lower Cretaceous sandstones as well as several types of Jurassic limestone have more probably been transported into the region initially by glacial action. These suggest that the exterior of the wall is not original but that it has been rebuilt. The flint facework both in the wall and the adjoining tower tends to be composed of semi-knapped to knapped, randomly dispersed and widely spaced flints, built in a style not unlike much of the Victorian flint walls elsewhere in the town. Nipples are, however, abundant and often as many as 0.5 in number per flint (calculation based on number of nipples observed on a count of 20 flints). Town Wall Road falls in elevation towards the east. Inside the wall the change in elevation is less significant, so that some evidence of a much modified rampire remains as the North East Tower is approached.

Only the relics of the lower part of the North East Tower exist. Access to the North East Tower (view only) and the inner wall face of stretch B2 may be gained by means of Town Wall Mews and Eden Place. In both localities the wall exposed has been much altered (to include 'modern' bricks) and partly rendered. In 1884, for much of its length, this stretch of wall was adjoined by buildings (Ordnance Survey 1884 plan). From Town Wall Mews there is some evidence of the wall walk (about 2.5m. above ground level) and the trace of one possible vaulted firing bay. Old bricks make up only a small proportion (about 5 per cent) of the inner wall and these appear to date from both the 15th (Type 3) and the 16th centuries (Type 7). More abundant are worn beach flint cobbles, some of which are broken. These occur together with a small proportion of field-picked flints. Rather less is visible from Eden Place, although very occasional fragmentary Types 3 and 7 bricks have been observed in the wall surface.

Figure 3.7 The exterior of wall stretch 'B2', viewed from Ferrier Road, displays highly modified, irregularly spaced, arrow/gun slits high in the wall. The shade changes in the flint wall facing are as much to do with pointing applied at different periods as to unlike building dates.

The ground externally, seen from the area of Ferrier Road, is somewhat lower than the surface inside the wall, suggesting that the walls here were at one-time rampired (Figure 3.7). Twenty, greatly modified, slightly irregularly spaced, arrow or gun slits can be seen high in the outside of the wall. These appear to be placed at the correct height to correspond with the one arch trace seen internally. Stretch 'B2' is broken by a number of now bricked-up doorways. Most appear 'modern', but one constructed in a Gothic style appears to be completed in Type 7 bricks. About 2m. south of the North East Tower, a few, fragmentary, Type 2 bricks set in concrete project from the foot of the wall. The outer wall face is finished in mainly broken (to knapped), relatively unworn, beach flint cobbles. A small proportion of the flints appear to have been quarried. These are widely spaced and set randomly; the conchoidally fractured flint surfaces only rarely show 'nipples' (in some areas none at all). Lumps of exotic rocks occur frequently in this wall (Table 4.3), and part of the stretch has been rebuilt in brick in an unusual double header version of 'rat-trap' bond almost certainly providing a date of about 1850 (Figure 3.8). A deeply inset blocked doorway towards the south end of this wall is probably framed externally in Type 7 bricks, the inset walls, however, have been constructed in bricks which would appear to date only from the 1830 to 1850 period. As with many of the flint faced walls, modern pointing applied at different periods tends to give the false impression of diverse building dates (Figure 3.7).

CHAPTER 3 THE TOWN WALL: 'A MODERN GEOLOGICAL PERLUSTRATION'

Figure 3.8 'Rat-trap' bonding seen in wall stretch 'B2' where the wall has been replaced with brickwork. This unusual, double header version almost certainly dates from about 1850.

Figure 3.9 The flint facework of wall stretch 'B3' just west of King Henry's Tower, where both patching and pointing add to the complexity of interpretation. This wall stretch facing Ferrier Road clearly shows marked changes in its different styles of flint dressing.

Figure 3.10 Detail of part of the wall face shown in Figure 3.9. At least two (possibly three) different styles are visible in this small surface area. To the left, the flints carry numerous nipples; to the bottom right, the more rounded flint cobbles have been broken differently to present a single flat to conchoidal face.

The small rectangular bastion at the western end of stretch 'B3' may originally have possessed quoins of Caen Stone. In the quoin nearest to wall 'B2' this stone has partially been replaced with stones of granite, and still more recently by blocks of concrete. Other quoins are now constructed of bricks and flints. A doorway into the bastion from St Nicholas churchyard is built of Type 7 bricks.

The lower part of the inner face of wall stretch 'B3' is obscured by a rampire (see Figure 5.4). The upper part shows considerable geological variety, this, the result of the recent insertion of whole and broken ashlars, some paving setts, and numerous, mainly broken, grave and memorial stones (the most recent of the readable dates being 1834). Widely and randomly spaced flints on this inner face are of whole to knapped flint beach cobbles, beach pebbles and field-picked varieties. Other exotic rock types are numerous (Table 4.3). Whilst no defensive wall walk or arched firing bays survive they are unlikely to have been absent in the earlier history of the wall. The wall core is revealed in a small section of the inner wall where a few fragments of Chalk are visible and the fragmentary exposed brick material proves difficult to date. On the external surface of this wall, a very small section of the wall facework has collapsed to expose the wall core. Only Type 7 bricks and mortar are revealed.

As with many stretches of outside wall, 'B3' exhibits scaffolding putlock holes (in this wall at two levels). All examined at Great Yarmouth appear to have been inserted late in the history of the wall's construction, and in the wall's towers many have been altered to include 'anti-roosting tiles' (see Ayers and Smith, 1988, 191) or bricks, to prevent pigeons nesting.

The flint facing in wall 'B3' is particularly complex. The various periods and patches of flintwork and pointing contribute to the complexity (Figures 3.9 and 3.10). Flint beach cobbles are in places largely broken, but they can elsewhere be knapped and in small patches poorly squared. Low in the wall, galleting occurs in a very small area part way between the Ferrier Road children's playground and King Henry's Tower. In some areas the flints are well coursed, in others randomly placed. Nipple counts vary from none to as many as 0.8 per flint (based on 20 flint face counts). Both Caen Stone and a few pieces of Barnack Stone are included in the wall face but other exotic stone types are generally absent.

At the east end of wall 'B3' is King Henry's Tower (said to be named after King Henry III {Ecclestone and Ecclestone, 1959, 90}), the only octagonal tower in the walls (Figure 3.11). It is 9.65m. wide between walls and Turner (1970, 144) indicates that it now stands only 3.66m. high. Access to the tower's interior could be by means of a locked gate, this gate possibly occupying the position of the original wall walk entry point from the west. Viewed from this gate, bricks in fragmentary form comprise only about 20 per cent of the interior wall structure: part of the remainder being of broken and unbroken flint cobbles. The lowest visible course at the base of the tower is of flint cobbles, this is followed by three courses of Caen Stone which are followed by flint external facing stones as elsewhere in the walls. The

Figure 3.11 A view from the east of the octagonal King Henry's Tower and St Nicholas Church.
From the tower the wall originally ran due south; it was removed in order to extend the graveyard towards the east

Figure 3.12 These randomly placed and well spaced, broken, rounded, flint beach cobbles are set into part of the south-east wall of King Henry's Tower. Nipples are uncommon – but a couple is present on the bottom left flints. This workmanship is thought to be of Victorian origin (see section *4.4.2*).

quoins of the tower seem to have originally been of Caen Stone, but this has been replaced in part by a fairly soft deltaic sandstone which might well have been extracted from the Middle Jurassic, Ravenscar Group of Yorkshire. The sandstone has, in turn, been partly replaced by blocks of Middle Jurassic Inferior Oolite. The external flint facework of the tower continues in the variable form of the wall immediately to the west, but it includes a significant area of widely spaced, broken to knapped, rounded beach cobbles (Figure 3.12). A few bricks are randomly included in the northern walls, probably as replacements, at least one part brick probably being of category Type 2. Palmer (1864a, 121) stated that the tower was used to deposit bones in at the time, subsequent to the Reformation, when the charnel house was destroyed.

A square King Henry's Tower is illustrated on the medieval enceinte (enclosure) of the town of Great Yarmouth, prepared in 1588 in anticipation of a possible Spanish invasion (see Frontispiece). The illustration was described by O'Neil and Stephens (1942). This 1588 document also displays mounts or ravelins that may have been planned but are not known to have been built. Clearly it must be treated with caution and there can be no certainty that the tower was ever other than octagonal in outline.

3.2.3 Stretch 'C', King Henry's Tower to Fishers Court

The northern part of stretch C is only represented by the remains of its rampire; to the south of this the wall may be divided into:-
'C1' – the wall running parallel to Priory Gardens; this being followed by a further wall gap once occupied in part by the Pudding Gate, and to the south of this,

CHAPTER 3 THE TOWN WALL: 'A MODERN GEOLOGICAL PERLUSTRATION'

Figure 3.13 Evidence of a one-time rampire is obvious in this portion of stretch 'C1' where in the grounds of St Nicholas church the town wall has been removed to enlarge the churchyard. St Nicholas Church viewed here from the north-east and extensively rebuilt after bombing in the last war, stands in the background.

'C2' – the area bordering the north of St Nicholas School and situated to the north of Fishers Court.

The early town walls no doubt skirted the original church and its graveyard, but St Nicholas churchyard now extends both to the east and west of the old wall line to the south of King Henry's Tower. Now, only the line of the smoothed out rampire can be determined (Figure 3.13). Somewhere in this area of the site of the wall St Nicholas Gate once existed. It is stated that the gate was blocked up at the time of the Reformation. Together with its tower, it was demolished (or partly demolished) in 1642 (Swinden, 1772, 100; Palmer, 1864a, 121), and the site finally cleared in 1799.

The northern extremity of 'C1' wall is marked by a small, apparently solid, turret. This might possibly mark the original boundary of the wall of the precincts of the Benedictine Priory built to the south-east of the early St Nicholas Church (see Palmer, 1864a, 120)(Figure 3.14). A further similar turret to the south may mark the southern limit. Access to the inside of the partly rampired wall is readily possible via the Priory Nursery Centre and between 6 and 7, Priory Gardens. At both localities the wall has been extensively repaired and rendered, but at the first locality likely evidence of the former wall sentry walk is present.

The exterior of the wall is adjoined by a footpath. In the northern portion, the flints are sometimes of boulder dimensions (and up to about 300mm. in size); their cortex and limited degree of weathering would be consistent with their acquisition from an Upper Chalk wave-cut platform (see Section *2.3.1*). Carter (1980b, 303), believed that these large flints gave this wall a unique appearance. The flints are generally broken; they are widely spaced and laid in poorly-delineated courses. The

Figure 3.14 A Victorian engraving of the hall of the Benedictine Priory which was restored in 1852 and still stands. This engraving was reproduced from Palmer (1864a).

flint faces bear conchoidal fractures and a few to a moderate number of nipples (up to 0.4 per flint based on

35

Figure 3.15 A boulder of slightly gneissose granite, with dark bands rich in the minerals biotite and hornblende. This broken boulder was set into the external flintwork of the southern end of wall stretch 'C1'. The boulder had, no doubt, originally served as ballast and may have in the first instance been gathered from the Baltic region.

Figure 3.17 This flint cobble, which is set into the facework of stretch 'C2', shows marked striations on its chalky external cortex. These glacial striae (parallel to the ruler) have been caused by the movement of the flint in the past within an ice sheet.

Figure 3.16 Set into the flintwork face of wall stretch 'C1', this large igneous boulder was broken in two or more pieces. It had probably initially served as part of a ship's ballast. The rock is a diorite (dolerite) which has subsequently weathered and exfoliated.

Figure 3.18 In stretch 'C2' a portion of the town wall has recently been removed to permit more light for St Nicholas School. The bricks of the wall core exposed are mainly of Type 2, but Type 3 bricks are also, present. The top of the Type 3 brick illustrated here reveals a double-margin or sunken margin, a feature which is distinctive but difficult to create or explain.

20 flint counts). This wall possesses many included different exotic rocks (see Table 4.3), including boulders of granite and exfoliated dolerite (Figures 3.15 and 3.16). Many of the large water-worn cobbles in the wall face may once have served as ships' ballast.

Moving to the southern end of 'C1', the external flintwork becomes initially better coursed. There are traces of a few highly modified arrow slits or gun ports and some putlock holes, but otherwise the wall is not unlike the northern portion. The upper portions of the wall show evidence of lifts, or being built in stages (see Figure 2.7). The wall at its southern extremity, however, shows areas of markedly different and varied styles of flint facework, including a small patch of galleting. Although two reconstructed arrow slits are present in this portion of the wall, much of the flint facework appears to date from the 18th, or more probably the 19th century. Nowhere was the core of wall 'C1' fully exposed.

Where the roads Priory Plain and St Nicholas Road exist, Swinden's twelfth tower and Pudding Gate once stood (Swinden, 1772). The gate was demolished in 1837 (Palmer, 1864a, 120; Ecclestone and Ecclestone, 1959, 90). An engraving of the gate was published in *Pictures of Old Yarmouth* (1897). The wall (within a small private car park) comprising approximately the first 20m. to the south of these thoroughfares forms stretch 'C2' of this analysis. The flint face shares some of the characteristics exhibited by stretch 'C1'. In particular, some of the broken to knapped flints are of boulder size and exotic rocks are of similar, not insignificant, abundance (see Table 4.3). A glacially striated cortex-surfaced flint is present in the face (Figure 3.17).

Towards the southern end of the C2' stretch, the wall

CHAPTER 3 THE TOWN WALL: 'A MODERN GEOLOGICAL PERLUSTRATION'

Figure 3.19 The view facing north-east from St Nicholas School of the inside of the wall and its firing bays. These cover the northern part of stretch '*D1*' extending into stretch '*C2*'. The wall has been capped with modern bricks at the height of the sentry-walk.

was fairly recently broken through to provide light for the St Nicholas School. The wall core is probably exposed better here than elsewhere and its composition is quite unlike anywhere else seen in the entire wall circuit. It was initially thought that all the bricks were of Type 2, the thick 1300-1350 brick. In occurrence they are here less fragmented (although complete bricks remain rare). The brick surfaces reflect drying between straw (Figure 2.8). It appeared probable that this was the only place in the town walls where a wall of the early 14th century was preserved and visible. On the occasion of a re-examination, however, a few Type 3 bricks were also noted. On two of these bricks the unusual, and difficult to explain, 'double-margin' or 'sunken margin' (Firman and Firman, 1983), could be observed (Figure 3.18); a feature in this brick type that from studies elsewhere is thought might help to confirm a 1430 to 1450 date. There is evidence at this locality that the present outer flint wall face has been rebuilt against the wall core.

3. 2. 4 Stretch 'D', St Nicholas School, Hospital Tower, Market Gate

The Hospital Tower divides this wall stretch into north, '*D1*', and south, '*D2*', sections.

The inner face of stretch '*D1*' (extending in to stretch '*C2*') displays an arcade of 16 arched firing bays which occur in the grounds of St Nicholas School; all but the southernmost two support the sentry-walk and a parapet. This provides one of the best preserved areas of the inner face of the wall in the entire circuit (Figure 3.19). The northernmost piers are built almost entirely of Type 2

Figure 3.20 Two small micaceous sandstone seats have been inserted into the walls of this firing bay seen from the grounds of St Nicholas School. Although its outline is poorly visible, the arrow slit in the bay, the sixth to the south of St Nicholas Road, has been bricked up. The same bay is visible in Figure 3.19.

Figure 3.21 The view of the exterior of the wall, viewed towards the north-north-west, from Fishers Court, stretch '*D1*'. Here, much of the flint facework has fallen away, to expose a partially repaired, wall core. Type 2 bricks are common in the original core in which the brick to mortar ratio for the wall is unusually high.

Figures 3.22 The author holds a tape beside (and to the right of) the line of a small fracture or displacement in part of the core fabric shown in Figure 3.21. This fracture is thought to pre-date the external flintwork, for where this is present it is not displaced. A further example of a similar fracture may be observed beside the tape in the figure on the right.

CHAPTER 3 THE TOWN WALL: 'A MODERN GEOLOGICAL PERLUSTRATION'

and nipples are rare, but exotic material is common (up to 20 per cent of the face) in certain patches (Table 4.3). Twelve arrow slits are preserved in the southern external wall of *'D1'* and their height in the wall is clear evidence of the internal rampire. Though the exposed wall core has been repaired, the original bricks are principally of the early Type 2, with fragments of Types 1, 3 and 4 bricks also present (Table 4.1). Bricks in the core are disposed more or less randomly or only very roughly coursed.

Hospital Tower (also known on occasions as St Mary's Tower, Turner, 1970, 144) which is named after the once adjoining medieval Hospital of St Mary, possesses squinch arches opening on to the wall walk (Figure 3.23). The face of the tower is made up of a mixture of well spaced, knapped flint cobbles which have been both quarried and, gathered from the proximity of a Chalk beach. Nipples are uncommon and there is a quantity of included exotics (Table 4.3). In the lower portion of the tower's walls the flints are set randomly but higher they become poorly coursed. In 1864, Palmer (1864a, 120) described the tower as possessing a modern roof.

The inner face of wall *'D2'* to the south of the tower can be seen from the Dissenters' cemetery, access to which may be gained, with permission, via the school. A further eleven, partially reconstructed, vaulted firing bays, all with associated sentry-walk, are preserved. Near the Hospital Tower the bays have been reconstructed in an imported Type 5 'clinker' yellow brick probably made in the late 15th or early 16th century. These bricks were in all probability made in Holland, a less likely source being the Baltic region. Although similar bricks have been seen in King's Lynn, no bricks of this type were observed elsewhere in the Great Yarmouth walls. The arrow slits in the northern part of this wall have only been bricked up in the last 50 years.

Although Palmer (1864a, 120) suggested otherwise, the difference in the wall elevation between the two sides of *'D2'*, clearly emphasises the presence of the internal rampire. The outside of *'D2'* wall is less well exposed than *'D1'*. In the 1884 Ordnance Survey plan, over its entire length, there were enclosing buildings and evidence of these buildings is clear (Figure 3.24). Presumably for this reason exotic rock types are extremely rare. The typical flint facework is mainly of large beach cobbles, these are moderately spaced, knapped or squared, with a moderate number of nipples (about 0.2 nipples per flint based on 20 flint counts), and poorly coursed. In a small area about half way towards the southern extremity of this wall stretch the facing has fallen to expose the wall core. About 70 per cent of the core material appears to be of fragmentary bricks; much of this is of Type 2 bricks, with some Type 3 and Type 4 material (see Table 4.1).

Market Gate, and probably the wall immediately to its north, is stated as having been demolished in 1830. The top of the gate had been previously removed in 1797 (Palmer, 1864a, 119).

Figure 3.23 Hospital Tower, viewed from Fishers Court to the north. A squinch arch carries the wall walk to a door in the tower. The ground is noticeably lower on the east side (outside) of the wall.

bricks and may well be original, although the proportion of these bricks compared with early 16th century (Type 6) bricks in the archwork probably reflects the extent of rebuilding. A few re-used fragmentary Type 1 bricks occur in the eleventh arch from St Nicholas road and in the sixth bay two small 'seats', of khaki, micaceous and glauconitic sandstone with moderately rounded grains, have been inserted into the brickwork (Figure 3.20). Based on a continuing and similar distribution of bricks within the archwork and piers as observed at the inner face, the minimum number of bricks used per arch and pier was here estimated to be of the order of 1,200.

The outer face of wall *'D1'* may be seen from Fishers Court. Much of the flint facework has fallen away to expose the wall core (Figure 3.21). Much (possibly all) of the facework has been repaired, in some instances with modern brickwork. In support of the flint facework being of later date, there are several instances where the wall core is fractured through slight movement, but the flint face is not displaced and remains unbroken (Figures 3.22). Patches are variously faced with broken to well squared, flint cobbles, either that have been quarried or from a Chalk beach environment. There is no galleting

The Medieval Town Wall of Great Yarmouth, Norfolk, U.K.

Figure 3.24 Wall stretch 'D2' and Hospital Tower viewed from the south-east. This external face to the wall is partially enclosed in buildings and in 1884 the wall was completely faced with buildings (in part an abattoir). 'Exotic' rocks are absent in this wall face presumably because it was not exposed over more recent centuries. A new wall has been erected above sentry-walk level.

Figure 3.25 A view southwards along stretch 'E' of the wall at the northern end of the Market Gates Shopping Centre showing clearly the archwork to the firing bays.

Figure 3.26 Detail of the part of the arch of the fourth bay south of Market Gates road partly figured in Figure 3.25. The edge of the arch to the firing bay is constructed of brick headers and stretchers placed alternately in pairs. The bricks are largely of Type 3 and of mid 15th century age. The diameter of the lens cap measures 50mm.

CHAPTER 3 THE TOWN WALL: 'A MODERN GEOLOGICAL PERLUSTRATION'

3.2.5 Stretch 'E', Market Gates Shopping Area

Palmer (1864a, 108), citing Manship (1619) stated that about 1588, the upper parts of the wall between Market Gate (stretch 'E') and Blackfriars Tower (at the southern end of stretch 'L') were reconstructed in response to the threats of the Spanish Armada. This statement will be examined in each of the ensuing accounts. Manship (1619, 73) actually indicated that the walls over this stretch were rampired during 1587, with the portion at Blackfriars being further rampired in 1588 (p. 74).

Dwarfed by the 'not at all sympathetic' (Pevsner and Wilson, 1997, 520) Market Gates Shopping Centre the wall remains remarkably well preserved, if much more modified than the stretch seen at 'D1'. Access, at either end, is from a path leading to the public toilets. For much of this stretch the wall can be viewed on either side and the evidence of the rampire on the inner side is unambiguous (see also Emery, 1998, 1). The upper portion of the wall, especially above sentry walk level has been levelled and modified in recent times. A raised area 'encompassed by a wall, higher than the Town Wall', and on the inside of the wall, generally referred to as the 'Main Guard', occurred in this area according to Palmer (1864a, 119) (See section **5.4**).

Between the road to the north and Guard Tower, seventeen vaulted firing bays each with an arrow slit are clearly exposed (Figure 3.25). The arrow slits have been constructed from Type 2 and 3 bricks. Five further arrow slits are visible to the south of the 'D' shaped Guard Tower, but their inner archwork is concealed by modern concrete (as is the southernmost arch to the north of the tower). Inside widths of the northern nine arches varied from a minimum of 2.00m. to 2.26m., the unusually narrow pier widths (see section **2.4**) correspondingly varying from 0.65m. to 0.73m. Bricks lining the arches at the inner wall surface were often placed with headers and stretchers alternating in Type 3, 15th century bricks (Figure 3.26). But the brick compositions of different arches and their inner linings were apt to vary with Type 2, some Type 4 and late Type 3, and perhaps most commonly Type 6/7 bricks. It would appear that these vaulted arches were built, or more probably rebuilt, in the first half of the 16th century, in part at least reusing earlier materials. A few of the firing bays exhibit relieving arches built of fragmentary Type 2 bricks. Swinden (1772, 88), referring to the murage records of 1342, described the purchase of lime made to whiten the walls. He recorded an Adam de Melton providing 4½ treys of quicklime for a cost of 4s. 4½d. This quicklime must have presumably been used for the inside of the towers or for the firing bays. The present author was unable to confirm, as original, any lime wash seen in either the arches or elsewhere.

On the outer wall of stretch 'E', which carries various putlock holes, the flints with fair to moderate cortex were poorly coursed, but better knapped to the north of Guard Tower than to the south. In both areas nipples were rare (about 0.1nipples per flint based on a 20 flint count, for a moderately squared patch) and galleting absent. Exotic rock inclusions were possibly absent in the north until a few metres from the tower (this wall was built against and hidden in the 1884 Ordnance Survey plan); those rock types recorded in Table 4.3 being observed in the area close to the tower.

The rib-vaulted squinch arches which carry the wall walk into Guard Tower from the north and south are well preserved (Figure 3.27), although the rib-vaulting can now only be seen on the south side. The south squinch arch appears to be preserved only in Type 2 bricks (Figure 3.28; but seen only at a distance). If this observation is correct the tower, although now much altered, must have been erected initially in the first half of the 14th century. The lower part of the tower is also well preserved and displays two rows of five arrow slits, the lower five being aligned with the wall slits. The flint faces, both in the tower and the wall to its south, vary at different levels through their height in the quality of their coursework. Flints at some levels in the tower carry rather more nipples. At no place in stretch 'E' is the wall core visible.

Figure 3.27 Guard Tower, set in the Market Gates Shopping precinct and viewed from the south-east.

A significant gap exists in the presence of the wall between stretches 'E' (to the south of Theatre Plain) and 'F.' This is in part currently occupied by Regent Street

Figure 3.28 The detail of the underside of the squinch arch visible in Figure 3.27 present on the south side of Guard Tower. From distant observation the bricks used to create the intricate rib-vaulting appear of Type 2, the craftsmanship, therefore, being particularly elaborate for the mid 15th century.

but in the past Swinden's eighth tower, to which a new roof is recorded as being added in 1542 (Rutledge and Rickwood, 1970, 49), and Oxney's Gate stood within this area. Oxney's Gate is said to have been demolished in 1766 (Palmer, 1864a, 118).

3.2.6 Stretch 'F', Regent Street, Pinnacle Tower to Shave's Tower

A plaque indicating the date of the demolition of Oxney's Gate is displayed on the south side of Regent Street, the wall to which it is fixed, however, appears relatively modern. From a small back room in British Home Stores, to the south of Regent Street, it is possible to gain access to two vaulted firing bays on the inside of the wall. This locality is very close to the one (now beneath the British Home Stores fire escape) at which Green (1970), in 1955, widened a small square test trench to try unsuccessfully to examine the base of the inside of the wall. In this excavation, beneath the soils of the rampire, he discovered 'odd small sherds of fourteenth-century type pottery' (p. 111, layer 2), possibly confirming the date for the original erection of the wall. From the details provided by Green it is obvious that the current inside exposure of the wall is a copy constructed in the 1955-1960 period to the recognised wall style. The bricks used are believed to confirm this observation.

The outer face of the same wall is more extensively exposed at a lower level (but to the south of the British Home Stores firing bay site) behind the various surgeries on the west side of Alexandra Road. Close to the wall

Figure 3.29 Wall stretch 'F', viewed from a point behind No. 4, Alexander Road, very close to the place where Green (1970) described an excavation just outside the wall. The view towards the north, part way along the wall, shows a marked change in the style of the flintwork facing, with the nearer work the more recent (see section *4.4.2*). A visible vent in the wall would suggest a void, probably into a firing bay, but those occupying the building inside the wall know of no access to this basement area.

face behind No.4, and opposite the test trench inside the wall, Green (1970, 113), in July 1955, dug another small trench and examined the exterior of the wall to its base (Figure 2.2). Below miscellaneous 19th century debris, he discovered several undated early sherds (layer 4), and deeper in layers 5 to 8, late 17th century materials. At the wall base, below the water table, and discovered 'by probing', the wall was thought to rest on flagstones. The flint facework in the vicinity of the test pit is of whole to semi-knapped, moderately spaced, well coursed flint cobbles from a Chalk beach. Some flints carry nipples and putlock holes are present (Figure 3.29). In the higher wall, which from the evidence of the wall interior has obviously been rebuilt, exotic boulders are evident.

There would have been about eleven firing bays in this stretch of wall 'F' and Pinnacle Tower to the south. The arrow slits of some of these remain. These, where early, are constructed of later Type 3 bricks with rarely included pieces of Type 2 and 4. Some of the slits have been filled with sheet plastic, indicating the presence of firing bays behind the face. Investigation suggests that these bays are not accessible and occur below the solid floor of British Home Stores. As Pinnacle Tower is approached the flints tend to become larger, more widely spaced and random in their distribution. An unusual *Polydora*-bored beach boulder of basalt is present in the face.

Figure 3.30 Pinnacle Tower as observed from the north shows the remnants of a squinch arch from an original wall walk. The tower has been significantly altered and the flat inside wall completely rebuilt.

Pinnacle Tower (Figure 3.30) was, when visited, a pigeon roost and in poor repair. In the 17th to 18th century it was used as a lock-up and a stable (Palmer, 1864a, 118). It is situated behind the residential home in Alexandra Road. It is again 'D' shaped and its name is no doubt derived from its conical roof (Wilton, 1979, 32). This roof was probably added in 1542 (*Norfolk Record Office* Y/C18/6, fo 29v.). The open-work weather vane that crowns the tower bears the date 1680 and the initials of the then bailiffs and chamberlains (Palmer, 1864a, 118). At its base, Turner (1970, 143) states the walls of the tower are 1.17m. thick. The outer face of the tower possesses similar flintwork to the wall to the immediate north, although above walkway height it is better coursed, perhaps indicative of later workmanship. The tower has positions for five arrow slits at the same height as the adjoining wall slits, the easternmost having been replaced by a Victorian sash window. A squinch arch is preserved in Type 7 bricks on the north side of the tower, whilst the inside flat wall of the tower has been replaced entirely with relatively modern bricks. No wall core is visible in stretch 'F'. The wall between Pinnacle Tower and Shave's Tower no longer stands.

3.2.7 Stretch 'G', Shave's Tower, New Gate and the East Mount

Only two parts of this stretch of the wall remain; '*G1*' representing what remains of Shave's Tower, and '*G2*', the East Mount, a later addition to the defences.

Only a small part of Shave's Tower, '*G1*' survives. Rutledge and Rickwood (1970, 46; citing *Great Yarmouth Assembly Minutes*, 1538-1545, fo 23r.), records the tower in the wall against Schave's (*sic*) dwelling place being repaired on 18th August 1542. The south-facing outside wall of the tower that can be examined is somewhat repaired. It is faced with whole to knapped, well spaced, flint cobbles from a Chalk beach. The broken faces bear some nipples. Although the flints are randomly placed in the lower part of the tower they become moderately coursed higher in the wall. Putlock holes are present and exotic blocks are common (Table 4.3). A patch of mid to late 15th century brickwork is present in the wall.

New Gate (also recorded as Chapel Gate and St George's Gate) was described by Palmer (1864a, 117) as having been erected in the time of Elizabeth I (possibly 1601) and demolished in 1776. This date of demolition is recorded on the commemorative wall plaque. Palmer (1864a, 117-118) records that in 1789, 1.8m. of wall on the south side of the site of New Gate was taken down. The church of St George, at the time of writing a disused theatre, is said to have been completed in 1714 (Palmer, 1864a, 116). Pevsner and Wilson (1997, 498) indicate that construction commenced on this date and the date of completion was 1716. If either of these dates is correct the church cannot stand on the site of the gate, and it probably stands to the west of the line of the wall.

Figure 3.31 To the rear of the terrace of houses facing on to St Georges Road (stretch 'G') the East Mount walling is just visible, with access only via the houses. The wall is clearly battered and made of Type 7 bricks with interspersed blocks of ashlar Caen Stone.

The construction of the East Mount, 'G2', is thought to have begun in 1569. Though triangular in shape, it does not appear to have been a true ravelin (see O'Neil and Stephens, 1942, 2; Pevsner and Wilson, 1997, 519), for it remains connected to the inner walled town, the earlier town wall having been broken through in its construction. In 1569 (and possibly 1570), Manship (1619, 47) took part as a schoolboy in assisting with wall repairs to create 'a very high mount of earth'. In 1588, he records the enclosing of the lower parts of the same walls with 'brick and free-stone' (see also Swinden, 1772, 97-98). Manship (1619, 40) indicated that some of the material used came from the old charnel house which was pulled down in the same year. Palmer (1864a, 107-108) stated that the structure was breasted with flags: he also records (p. 117) that as late as 1620 material from the castle was used for the walls 'enclosing the mount'. The battered walls of the Mount can be observed immediately behind the houses facing St George's Road (Figure 3.31). The face is entirely built of Type 7 bricks, but liberally dispersed in it are variously sized and obviously reused blocks of ashlar Caen Stone (see section *2.3.4*d). The east end of the Mount may be observed, supporting a plaque, where the road turns into St Peter's Plain.

3.2.8 Stretch 'H', south of East Mount, Little Mount Gate and Harris's Tower

To the south of East Mount, although once covered by buildings related to the old Yarmouth Hospital, the outer face of the wall is well exposed behind Ravelin House on St Peter's Plain. Two vaulted firing bays (Figure 3.32) have here been broken into (now gated) from the outside of the wall (through about 0.5m. of wall). The inside of

Figure 3.32 Remarkably, to the west of Ravelin House and St Peter's Plain, stretch 'H', the outside east face of the wall, has been broken through to gain access to two firing bays (one only shown; Rose (1991, 201) has suggested that these access points were made possibly in the late 18th century when the owners of adjoining cottages broke through the wall to gain an extra cellar). The entrances are now gated and locked. Soil, in the lower levels of the internal rampire, had presumably to be removed to enable their use.

CHAPTER 3 THE TOWN WALL: 'A MODERN GEOLOGICAL PERLUSTRATION'

Figure 3.33 The outside of the town wall, stretch 'H', behind the mews to the west of St Peter's Plain. A putlock hole is visible in the flint facework. To the left, the face has fallen to expose the wall core.

the wall (Dene Side) was clearly rampired and these firing bays remain well below the level of the rampire. In the vicinity of the bays, bricks make up only perhaps 35 per cent of the core material (although rather more to the south), 35 per cent being flint cobbles and about 30 per cent mortar. The majority of the bricks are early 15th century Type 3, but fragmentary Types 2, 3, and 6/7 are included (Table 4.1).

Following a short gap covered by garages, the outside of the wall is again visible further to the south, again capped by the rear walls of buildings in Dene Side. Accessed by the mews behind St Peter's Plain, the wall, still fully rampired on the inside is less rendered. The flint face is formed of Chalk beach derived flints, with a few probably quarried. They are broken to knapped in most instances, but in a few areas they are squared with limited patches of galleting, and well coursed over most of the wall. Exotic boulders appear absent and nipples on the flint faces vary from none to a few. Putlock holes are present in the walls (Figure 3.33). Five, possibly six, infilled and part repaired arrow slits occur in the wall; Type 2 and Type 6/7 bricks (used in nearly equal proportions), together with a Type 3 and a few Type 4 bricks, having been used in their construction (Table 4.1). The wall core is partly exposed and consists of a range of similar brick types, these occurring with relatively small unbroken flint beach cobbles.

Figure 3.34 The stump of Harris's Tower, surmounted by an early 19th century house, as viewed from the north. The wall changes slightly in direction at this point. Palmer (1864a, 115) indicated that the ground floor of the tower had been used as a stable.

The wall cannot be observed in the area around York Road, once in part occupied by Little Mount Gate, a gate that was demolished in 1804 (Palmer, 1864a, 115-116). Ecclestone and Ecclestone (1959, 90) advise that a length of adjoining wall was also removed in the same year. The outside of the wall is next seen just to the north of the unusual stump of Harris's Tower (Figure 3.34). This stump is surmounted by a brick-built early 19th century house with its front in the higher rampired Dene Side. The tower is 'D' shaped and has been partially repaired. The outer face is of Chalk beach derived flint boulders that are broken to knapped, all being moderately spaced and poorly coursed. Both nipples and replacement exotic boulders are present. The core of the tower wall reveals bricks, mainly of fragmentary Types 2, 3 and 4, in order of abundance, all set in nearly the same quantity of mortar as brick (Table 4.2). Broken flint cobbles constitute only 15 to 20 per cent of the core. The tower displays a number of much altered arrow slits which are at ground floor level and the tower stump appears to have been levelled at the height of the wall walk. Finally, a short stretch to the south of the tower forms the party wall between adjacent buildings so that it is virtually obscured from view until the rebuilt southern end is exposed on the northern side of Lancaster Road.

3.2.9 Stretch 'J', White Lion Gate to Garden Gate (Alma Road)

The substantial gap in the wall to the south of Lancaster Road was in part occupied at one time by White Lion Gate. Also referred to as Ropemaker's Gate (Eccleston and Eccleston, 1959, 90), it is said to have been demolished in 1785 (Palmer, 1864a, 115). The Greek Church of St Spyridon, formerly St Peter, constructed in 1830, now occupies part of the site. Specifically for the church a length of wall immediately to the north of the gate was taken down about this time (1833).

About 40m. to the south of the church, the wall may again be observed between Dene Side and Blackfriars' Road. The inside of the wall is again rampired, the rampire rising nearly to the height of the sentry-walk so that only the vaulting of many firing bays is visible. There is some evidence of the battlement above the walk way and this has been frequently rebuilt (Figure 3.35). Just above the arches the wall core is variously exposed: all the bricks are fragmentary with total bricks and flint beach cobbles, some of which are broken, in approximately equal proportions (30 to 35 percent), and wall mortar predominating (up to 40 per cent). The number of included bricks tends to decrease towards the south over this 40m. of exposed wall stretch. Seventy per cent of the included brick fragments tend to be of Type 2 bricks, but with much of the other 30 per cent being of Type 6/7 bricks the wall appears to have been rebuilt no earlier than the mid 16th century (Table 4.1).

To the south, the inside of the wall passes behind a Community Services building and is temporarily obscured before being well exposed to its south (with nearly a further 60m. of inner wall present). In total, at least 26 arrow bays/slits may be examined in stretch 'J', despite the extent of the rampiring, which decreases in height as Garden Gate is approached (Figure 3.36). The area of more intensive rampire remains approximates to the position where according to Palmer (1864a, 114-115) Symonds' Seat Mount existed. This mount supported a more recently built look-out until it was pulled down a short time before 1864, and the mount subsequently, partially levelled. Various measurements on the firing bay arches where they were more fully exposed provided: widths of bay (Figure 2.11A), 2. 20 to 2. 27m.; width of pier (Figure 2.11B), 1.30m.; depth of bay (Figure 2.11C), 1.51m.; height of arrow slit, 1.02 to 1.22m.; width of arrow slit inclusive of embrasure 0.77 to 0.90m.; depth of arrow slit (equals outer wall thickness, Figure 2.11E), 0.49m. Figures 3.37 and 3.38 provide more detailed views of the 4th and 12th firing bays respectively, north of Garden Gate. Much of the wall carries both a repaired sentry-walk and castellated embattlements. It should be noted that the embrasures to the sentry-walk are very

Figure 3.35 A view from the north-west of the inside of the wall in stretch 'J' on Dene Side to the north of the Community Services building. Evidence of the rampire remains, and it still rises to almost cover the firing bays. Above the wall walk, which has been repaired with flint beach cobbles, the battlements have been extensively altered.

CHAPTER 3 THE TOWN WALL: 'A MODERN GEOLOGICAL PERLUSTRATION'

Figure 3.36 A similar view from the north-west of the inside of the wall to the south of the Community Services building in Dene Side (stretch 'J'). The firing bays become increasingly exposed proceeding towards the south and Garden Gate. In nearly all instances the bays have been removed to leave only their traces on the inside of the wall. The remnant bases of the firing bay piers remain (see also Figures 3.37 and 3.38).

unlikely to be original for they are unrelated in their positions to the arrow slits. The wall core towards the southern end of 'J' possesses a different distribution of brick types to that north of the Community Services building. Type 3 bricks at about 65 per cent of the total bricks predominate; Type 2 bricks make up about 30 per cent and Type 4 the remainder. The quantity of total brick material makes up about 40 per cent of the wall matrix, the mortar now noticeably includes Chalk debris and fragments of iron slag occur within the wall. An attempt to estimate the total brick usage per arch/pier, based on the fragmentary bricks visible, suggested that about 550 whole bricks might have been used. The arrow slits were variously reconstructed and infilled with bricks of Types 3 and 6/7 bricks and a small number of Type 4 bricks. The 4th and 5th arrow slits, respectively, are viewed from the exterior of the wall in Figures 3.39 and 3.40. The inner face of the wall where it is seen and unaltered is very similar to the wall core in composition.

With the exception of the northernmost 50m., the outer face of wall 'J' is continuously exposed. It may be examined from both the car park and the playground off Blackfriars' Road, near to the 'Time and Tide' Museum. The whole is variously patched (Figure 3.41), and using composition, mortar and styles an attempt to determine the age differences of the diverse areas was made (see section 4.4.2 for the details). The oldest flint facing in the wall appeared to be of coursed widely spaced flints, where the flints were broken to knapped and with their faces displaying a very frequent presence of nipples. Areas of coursed but knapped to squared flints, with galleting appeared to be younger, followed by areas of randomly placed broken flints in which nipples were absent. The last type of walling appeared to date from about Victorian times and it occasionally included areas of random included 'modern' bricks (Figure 3.41). At the top of the wall more recent work was completed with unbroken flint beach cobbles (see Figures 4.5 to 4.14 for further details of these flint face styles).

Exotic rocks which were probably replacements (Table 4.3) were found throughout all varieties of flint facing. The wall also includes areas of brick replacement, such as a, possibly early 19th century, doorway (with modern infilling bricks which had a likely local clay source similar to the composition of Type 2 bricks) towards the wall's southern end. The decorative cross inserted into the reconstructed wall's end facing onto Alma Road is made of Caen Stone (Figure 3.42).

3.2.10 Stretch 'K', Garden Gate to the South East Tower

Garden Gate, also known as Little Gate, was probably not built until 1636 (Palmer, 1864a, 114). Palmer (1864a, 114) and Pevsner and Wilson (1997, 517) place the date of demolition as 1776.

Where present, the short stretch of wall 'K' can only be examined on its outer face. The inside of the wall is adjoined by a number of buildings which front Trinity Place, including the pottery, all set at a somewhat higher elevation than the outside of the wall, indicative of the past presence of a rampire. In the northernmost portion of

Figure 3.37 The trace of the 4th firing bay arch to the north of Garden Gate. The bricks outlining the Gothic style arch are all of Type 2 bricks and the arch is probably of original mid 14th century construction. Inside the arch the wall has been rebuilt and it is of a later date.

Figure 3.38 Slightly more of this firing bay, the 12th bay north of Garden Gate, is preserved. Again, the arrow slit has been blocked. Most of the bricks are again of Type 2.

Figure 3.39 The arrow slit of the 4th firing bay north of Garden Gate viewed from the outside of the wall (compare with Figure 3.37). The slit has been completely replaced and infilled with bricks. These are displayed as headers of mainly Types 3, 6 and 7 bricks, but there are also a few of Type 4, and possibly at least one modern brick.

Figure 3.40 This figure displays the next arrow slit (the fifth) north of that shown in Figure 3.39. Set differently, but again completely replacing the original slit, this contains only various old bricks, and the replacement may have been done at the time of the rampiring in the mid 16th century.

Chapter 3 The Town Wall: 'A Modern Geological Perlustration'

Figure 3.41 At least three different styles of flint facework can be observed in this wall stretch 'J', opposite No. 7, Blackfriars' Road, near the Time and Tide Museum. Some of the workmanship is considered to be no older than Victorian.

Figure 3.42 A view of the wall towards the north from Garden Gate. Blackfriars' Road runs parallel to the wall on the right. The sentry-walk is preserved above the first few firing bays. The south, near end of the wall has been rebuilt and includes a Caen Stone cross presumably created from reused stone.

the stretch 'K' bounding the Jewish Cemetery, the wall has been rebuilt, with the following 10m. missing. South of this, in the area below the pottery, much of the face has been lost, damaged by, or in the removal of, the structures that once abutted the wall. Many of these have been demolished in relatively recent times (see Table 3.1). Where present the outer flintwork face appears to be oldest where it takes the form of well squared flints (Figure 3.43). It has been replaced by large Chalk beach cobbles of flint. Although the wall beneath the flint face is well exposed, unaltered wall core is difficult to find. The lowest 2m. of the wall below the pottery, for instance, has a new sandy cement mortar and numerous fragmentary horizontally placed bricks, and has apparently been rebuilt (Figure 3.44). Higher in the same wall, where the chalky mortar is more typical of earlier

Figure 3.43 The external face of the wall in stretch 'K', slightly to the south of the pottery. The moderately squared flints to the south (left) of this view are probably the oldest in this stretch of wall face. Where the flint facework has fallen the material in view has also been repaired, that is, it is not original wall core: note that it contains flint beach cobbles.

Figure 3.44 This exposed wall in which all trace of the flint facework has been lost, occurs outside the pottery in stretch 'K'. On first impression the lowest third (2m.) of the wall with numerous fragmentary bricks appears old, but the bricks are of different types and the wall contains abundant rounded beach cobbles. Higher, bricks are far less numerous and a chalky mortar forms much of the wall matrix. This probably represents the early wall core with bricks no younger than the mid 16th century.

building, bricks make up to only 30 per cent of the wall matrix and they are mainly of early to mid-16th century date but include a few Type 2, late Type 3 and Type 4 bricks (Table 4.1).

The outer wall face immediately north of the South East Tower retains modified battlements and includes boulders which probably reached Great Yarmouth as ships' ballast (Table 4.3).

3.2.11 Stretch 'L', South East Tower to Blackfriars' Tower

This portion of the Great Yarmouth walls can for convenience be subdivided into:
'L1' – South East Tower and the north-east to south-west adjacent wall.
'L2' – The north-south wall, north of the Charles Street pathway.
'L3' – The southernmost stretch of north-south wall to Blackfriars' Tower.

The 'D' shaped South East Tower is perhaps the most imposing tower to survive (Figures 3.45 to 3.47). It rises to 9.14m. (Turner, 1970, 143). According to Pevsner and Wilson (1997, 517) the lower stages of the tower possibly date from the early 14th century (but see below). The upper part, including the chequerwork, was probably constructed in the 16th century. It has been suggested that this may have occurred about 1512 (*Great Yarmouth Town Wall*, 1971, 4) when the town's fee-farm payments to the king were remitted (*Calendar of State Papers, 1509-1514*, 687). Further remissions were made later in the same century (see section *3.2.5*) and these modifications could well date to around the 1580s. The chequerwork, consisting of square flint-and-brick panels, four courses high, shows evidence of repair. Palmer (1864a, 114) referred to 'three curiously carved stone gurgoyles', only one of which appears to remain (Figure 3.48), and suggested that these came from the ruined Blackfriars' church. A gabled 19th century structure, described by Palmer (1864a, 114) as a 'small tenement' crowns the tower. Inserted, in the tower walls, presumably in the position of earlier arrow slits are gun

CHAPTER 3 THE TOWN WALL: 'A MODERN GEOLOGICAL PERLUSTRATION'

Figure 3.45 Great Yarmouth's most imposing tower is probably the South East Tower in Blackfriars' Road. Here the tower is viewed for the south-east. The four rows of chequerwork, although, like the tower, partially repaired, are believed to date from the 16th century and in this instance most of the visible, rounded beach cobbles are likely to be of this age. The wall break, created post 1904, is noticeable immediately this side of the tower.

Figure 3.46 A view very similar to that shown in Figure 3.45 is portrayed in this pen and ink drawing by Noel Spencer. The drawing is exhibited in the Time and Tide Museum and was an accession to the Great Yarmouth collections in 1983. Permission to reproduce this drawing is gratefully acknowledged.

loops. A few of these are of Caen Stone, but most have been replaced in recent centuries with Portland Stone, one at least being of the fossiliferous variety known as 'roach' (Figure 3.49).

Low in the outer face of the tower some of the flint facing is galleted and better squared with about 0.3 nipples per flint face. The majority of the lower face, however, tends to be of broken to knapped, somewhat rounded flint beach cobbles with limited cortex, in which nipples are only about 0.1 nipples per flint face. Low in the face, exotic boulders are relatively common (Table 4.3). Externally, the inside wall of the tower's 'D' is built of bricks; many of these are modern relating to repairs. Some, including one small patch, are of Type 2, but the majority are early to late 16th century Type 6/7 bricks. This would seem to negate the currently visible tower walls as being of a 14th century date.

Immediately to the south of the tower, in order to allow direct access between Trinity Place and Blackfriars'

Road, a cut has been created through the wall, this now creates an awkward relationship between the South East Tower and the northern end of wall '*L1*', leaving the trace of the most northerly arrow arch incomplete. The break in the wall is not referred to by Palmer (1864a, 114) nor is it shown on the Ordnance Survey plans of 1884 or 1904. It must, therefore, post-date the latter date.

The documented record of construction, rebuilding and general modification of stretch '*L*' is much fuller than that of any other part of the town's defences, a reflection of its particularly chequered history. Much of the historical record is to be seen in the geological composition of the various sections. Several events relate especially to stages of rebuilding in the 16th century:

a) The fire (1525) and dissolution (1539) of the local Blackfriars' monastery (see section *2.3.4*) which eventually (within 20 years, Rye, 1973, 498; citing Manship, 1619) provided a liberal source of stone. Rye's citation should be corrected for it has been on occasions repeated. Manship (1619, 38) actually wrote, 'the walls whereof, with the foundations, twenty-five years past were wholly digged up, and disposed to other uses'. (In the first instance, the error

51

Figure 3.47 This etching of the South East Tower from the north-west was executed by Mrs Bowyer Vaux and appeared in Palmer (1852, where it was incorrectly labelled the 'south west tower'). Considerable alterations to this view have occurred in the last 150 years.

Figure 3.49 This gun port on the South East Tower is constructed of 'roach' rock; a variety of Upper Jurassic, Portland Stone. High in the tower, similar gun ports are made of Caen Stone, for which, in the lower ports, the Portland Stone is probably a replacement.

Figure 3.48 Palmer (1864a, 114) referred to three stone 'gurgoyles' on the walls of the South East Tower. High on the walls, only one gargoyle, probably of Caen Stone, remains. It seems likely that these were reused from the ruined Blackfriars' Church.

probably arose when Palmer {1852, 379}, misquoted the period as 20 years).

b) The collapse of the wall between the South East Tower and Friar's Lane in 1577 (Rye, 1973, 502). This wall collapse date was given as 1557 by Palmer (1852, 390; 1854, 417). Manship (1619, 88), advised that the town was inundated by 'a great rage', although it cannot be certain that he was referring to the same tidal surge event.

c) The requirement, about 1588, to modify the wall in response to the threat of the Spanish Armada (see section *3.2.5*).

d) Repairs to a stretch of wall in the south are cited as being necessary in 1551 (Ecclestone and Ecclestone, 1959, 47). Records of this work may also be found in the *Great Yarmouth Assembly Book*, 1550-1559, in 1551 (Y/C19/1, fo. 15r., 21st April; Y/C19/1, fo. 20r., 14th June) and in 1556 (Y/C19/1, fo. 146r., 28th February; Y/C19/1, fo. 147v., 13th March).

Although re-cycled ashlar material is much in evidence throughout stretch *'L'*, it is unevenly distributed. Thus, on the external surface of the wall, it is found plentifully above the height of about 1.7m. throughout the major part of *'L1'*, but it tends to be absent at the northernmost end, and extends down to ground level at the south end. Reused quatrefoil column sections are very evident in the wall in the southern portion of *'L1'* (Figure 3.50). On the outer wall face of *'L1'*, four gun ports, built in Portland Stone, similar to those in the South East Tower are preserved immediately to the south of the Friar's Lane

CHAPTER 3 THE TOWN WALL: 'A MODERN GEOLOGICAL PERLUSTRATION'

Figure 3.50 The external surface of wall stretch '*L1*'. This wall displays numerous reused blocks of stone from the earlier Blackfriars' site. These include quatrefoil column sections (top left) probably from the church, as well as window-framing stones now reversed to splay outwards as embrasures. The height of these embrasures gives a realistic idea of the original size of the internal rampire upon which any defensive cannons would stand.

Figure 3.51 Tightly packed squared flints with gallets, on the outer surface of part of wall stretch '*L1*'. The different styles of flint facework are discussed in section *4.4.2* . It follows, that if this is apparently the earliest flintwork style in the wall, which here dates from the 1550s, the style is also likely to have been created in the mid 16th century.

passage. The majority of the flint facework is constructed of knapped to squared, Chalk beach derived flint cobbles, many of which still preserve cortex; bonded with flint chips in galleting style (Figure 3.51), where nipples often tend to be absent. A small area of typical Victorian style flintwork occurs as a repair low in the wall towards its southern end. Reused blocks included in the wall are principally of Caen Stone, Barnack Stone (a few examples of both stone types showing evidence of red discoloration resulting from fire (Figure 3.52) and *Viviparus* limesone ('Purbeck Marble'. The last of these stones occurs normally as fragments from tombs and presumably they were originally polished. Palmer (1854, 418) stated that when reused stones have been removed from the wall, in several instances tracings of mouldings have been found on their reverse. Window-framing stones of both Caen and Barnack Stone probably from the early church, high in the wall, appear to have been reused for much the same as the original purpose, but the embrasures now splay outwards from the wall (Figure 3.50). These may possibly have protected canon emplacements at the time when the walls were fully rampired. On the inner wall, traces of the firing bays which include the gun ports occur, bay three from the north being 2.20m. wide. The arches are built in both Type 2 and Type 6 bricks. The wall's inner surface has been repaired but the level of the sentry-walk remains in evidence.

Between sections '*L1*' and '*L2*' the wall is awkwardly off-set. The changes in wall orientation involve a quoin which is of Caen Stone but set in an angle of other than

Figure 3.52 This block of Barnack Stone (centre) is darker on the right where it has been discoloured red by fire. Set in this stretch 'L1' of the outside of the wall, it is enclosed by three pieces of Caen Stone and flints with galleting. Both the Caen and the Barnack Stones have apparently been reused from the fire devastated ruins of the Blackfriars' monastery and church, when the wall was rebuilt in the mid 16th century.

Figure 3.53 An off-set occurs in the wall between stretches 'L1' and 'L2' and it is shown here viewed from the south-east. The south end of wall 'L1' terminates in a Caen Stone quoin constructed at other than a right angle, suggesting that it was cut for the purpose, and is not, therefore, reused.

ninety degrees (Figure 3.53). It would be exceptional, therefore, for these stones to have been reused. A brick arch for a drain also occurs in the off-set.

Wall 'L2' shows evidence of a rampire throughout its length; it also possesses seven visible arrow slits at its northern end. The outer flint face at the northern end of this wall is not unlike that of 'L1', possibly less well knapped, containing rather more, slightly worn, beach boulders and slightly better coursed. Nipple numbers are very variable, suggesting reuse, with different patches varying from none to 0.5 per flint for typical counts of 20 flints. Towards the south in the same wall, nipples vary between 0.3 and 0.5 per flint, the flints are here better squared and gallets are profusely in evidence. Exotic boulders are common both on the outside and the inside of the wall (Table 4.3 and Figure 3.54). A north-facing ashlar quoin of Caen Stone (of uncertain purpose) is present on the exterior of the wall. Sixteenth century bricks are particularly abundant on the inside of the wall and where the core is partially exposed. Fragments of Chalk and iron slag are also present within the wall core. Much of stretch 'L' shows evidence of a wall walk and parapet; these have been extensively repaired with unbroken beach cobbles of flint.

The line of the wall is broken by the Charles Street pathway (Figure 3.55), constructed some time prior to the Ordnance Survey 1884 plan (and 1864), but apparently subsequent to the date of the Faden map of 1797 (see Figure 4.3). The path is lined for a short length on its northern side with a modern wall of 'recently' imported, igneous and metamorphic, Scandinavian ballast boulders. Stretch 'L3' to the south of the path includes four interesting wall breaks at 6.0, 14.4, 26.3 and 38.6m. from the path; the first of these appears to be slightly off-set but this actually can be related to the absence of the flint wall face on the south side of the break (Figures 3.56 and 3.57). Each is not (or only poorly) keyed into the next section and is created with mid 16th century bricks (within the 14.4m. join a few included Flemish bricks). These wall joins appear to possibly represent divides between sections of wall of the same building period (as may be the case for the Caen Stone quoin referred to above). If this is correct it displays a very unusual building practice. The character of the wall and its matrix is otherwise similar to that of 'L2', exotic boulder repairs being particularly abundant (Table 4.3 and Figure 3.58). Where it is visible, the wall core (Figure 3.59) contains only about 15 to 25 per cent bricks. This is outnumbered by flint cobbles (30 to 40 per cent), with mortar comprising the remainder. The core also contains some Chalk fragments. The bricks are almost exclusively of Type 6/7 with rare included pieces of late 15th century Type 3 bricks. At the south end of this wall and on its inside, the traces of a firing bay (width 2.22m.) and its northern pier (width 1.05m.) are present. A small area of bricks in the outer face immediately to the north of Blackfriars' Tower is set into a hole in the wall, these bricks have frogs, are marked 'R8' and are of 19th to 20th century date, having been made in metal formers.

CHAPTER 3 THE TOWN WALL: 'A MODERN GEOLOGICAL PERLUSTRATION'

Figure 3.54 This large, rounded cobble of Jurassic muddy limestone occurs within the outer flint face of the wall in stretch 'L2'. The cobble shows evidence of its previous existence in a shallow marine environment in that it shows borings (here minute holes) by the marine organism *Polydora*. This suggests that the cobble was earlier used as ballast. The lens cap has a diameter of 50mm.

Figure 3.55 The wall is seen here in section at the Charles Street pathway, southern end, of stretch 'L2'. The flint wall face, in this case repaired, can be seen to be of limited thickness. The external face of the wall can also be seen to include numerous ashlar blocks from the earlier Blackfriars' site.

Figure 3.56 Charles Street pathway viewed from the southeast with wall stretch 'L3' in the left foreground. An unusual wall break, created with mid 16th century bricks is evident in the wall. To the south of this break the wall has lost its flint facing.

Figure 3.57 The wall break viewed in Figure 3.56 is seen here in more detail. The wall to the south (left) has lost the external flint facework to expose the wall core.

Figure 3.58 Inside the wall stretch '*L3*', just north of Blackfriars' Tower, exotic boulders and cobbles are common. That the wall has been much repaired is evident from the many rounded flint beach cobbles and the row of igneous and much altered sedimentary rocks, some of which may be altered tuffs, were almost certainly previously used as ships' ballast.

Figure 3.59 The wall core is exposed occasionally on the inside of wall stretch '*L3*'. Flints and mortar both exceed brick fragments in quantity and the wall also includes some Chalk fragments which can be seen in this figure. The fragmentary bricks are almost exclusively of Types 6 and 7. The steel ruler is 300mm. in length.

3. 2. 12 Stretch 'M', Blackfriars' Tower, Palmer's Tower and the South Mount

It is useful to break this east to west stretch of wall into several component parts:
'M1' – Blackfriars' Tower.
'M2' – the wall between Blackfriars' Tower and Palmer's Tower.
'M3' – Palmer's Tower and the South Mount.

Blackfriars' Tower carries certain matrix characteristics from both the adjoining north-south wall '*L3*', and the east-west wall '*M2*'. Murage records of the Edward III period (probably about 1337-1338) suggest that work was in progress on the tower and it is said to have been completed in 1342 (Swinden, 1772, 83-84, 89). In 1566 it was ordered to be repaired (Palmer, 1854, 417). As with the South East Tower the height is said to be 9.14m. (Turner, 1970, 142). The upper portion has three levels, rather than four, of flint-and-brick chequerwork in squares (Figures 3.60 to 3.62). These have been repaired in recent times. A passage was cut through the tower in 1807 to permit ready access 'to Black Friars' gardens' (Palmer, 1854, 417; 1864a, 112; *Great Yarmouth Town Wall*, 1971, 6). The tower is said to have been repaired in 1969 (*Great Yarmouth Town Wall*, 1971, 6). The outer face of Blackfriars' Tower (Figure 3.60) resembles wall '*M2*' for it almost entirely lacks flint gallets. Areas of the flintwork do, however, vary; the work is semi knapped to knapped, moderately to tightly spaced, and moderately to poorly coursed, with a few small areas containing some gallets; the whole suggesting a variety of periods of workmanship. The rear (inner wall) of the tower has brickwork similar in date and style to that of wall '*L3*', but much of it has been repaired in fairly recent times, such as many of the areas which have been faced with worn, unbroken beach cobbles of flint.

CHAPTER 3 THE TOWN WALL: 'A MODERN GEOLOGICAL PERLUSTRATION'

Figure 3.60 Blackfriars' Tower is here viewed from the southeast. It supports three chequerwork rows of ornamentation. The passage through the tower was made in 1807. Blackfriars' Tower, as other towers, has been extensively repaired over a variety of different periods.

Figure 3.62 This figure of the rear of Blackfriars' Tower, taken from the north, was first published by Tingey (1913, where it was incorrectly described as the South East Tower). The changes to the adjoining walls and their relationships to enclosing buildings in less than a century are dramatic.

Figure 3.61 Noel Spencer's pen and ink drawing reveals the extent of the alterations to buildings surrounding the tower over the last half century. The view is similar to that in Figure 3.60. The drawing, a Yarmouth Museums' accession in 1983, is displayed in the Time and Tide Museum, and to which permission to publish is gratefully acknowledged.

Wall '$M2$' extends from east to west, parallel to Mariners' Road and a footpath extending along its inner side. It shows traces of 18 firing bays and/or arrow slits before the footpath turns to the north away from the wall. Both sides of '$M2$' wall were completely enclosed by attached buildings over recent centuries (see Rye, 1973, 478), these being removed in recent decades to leave the wall greatly scarred. Evidence of any rampire is limited to a slight difference in elevation on opposite sides of the wall. The firing bays have been removed to preserve principally the outer wall skin (Figures 2.11, measurement E, and 3.63), which is in one instance about 0.95m. thick, but 0.90m. thick elsewhere in an area where the flint face is missing (Figure 2.11, measurement G). The first arch to the west of Blackfriars' Tower is 2.21m. wide (Figure 2.11, measurement A), the thirteenth 2.24m. The wall is much repaired but early core brickwork consists of up to 80 per cent of fragmentary (many 'half' bricks) Type 6/7 bricks, much of the remainder being later 15th century Type 3 bricks, with rare Type 4 and the very occasional piece of Type 2. According to the area, brick material makes up anywhere between 40 and 65 per cent of the wall matrix, with the remainder being mortar in excess of mainly broken, beach-worn flint cobbles.

Figure 3.63 Wall stretch '*M2*' is seen here from the east. The inside of the wall is greatly defaced; the firing bays having been removed and their presence can only be detected by the positions of the much altered arrow slits. This wall, over recent centuries, was enclosed by attached buildings and, it seems possible that the firing bays were removed at the time of rampiring in the 16th century.

Figure 3.64 This figure displays the outside of the wall shown in Figure 3.63, observed from the south-east. Although the evidence of rampiring internally is slight, the difference in ground levels on either side of the wall is obvious. The wall shows evidence of repairs, and particularly of those undertaken in the last century.

Figure 3.65 Palmer's Tower is here seen from the south-east. The wall projecting from it towards the south (left) is modern. Again, much altered, the tower may have once supported a windmill.

Figure 3.66 Noel Spencer's pen and ink drawing of Palmer's Tower, again kindly reproduced from the Time and Tide Museum's, 1983 acquisition, collection, is a view drawn about 1955, from slightly more to the south than that shown in Figure 3.65.

CHAPTER 3 THE TOWN WALL: 'A MODERN GEOLOGICAL PERLUSTRATION'

Figure 3.67 This etching of the outside of the South Gate was prepared for Dawson Turner, the East Anglian 19th century historian. Although the North and South Gates to the town were the same width (a gate of 3.7m. set in a total width of 20.1m.; Swinden, 1772), they were somewhat different in style (compare with Figure 3.6). The figure is reproduced from Preston (1819).

The outside of wall '*M2*' has also been variously repaired (Figure 3.64). The flint facework is generally knapped, but in patches squared, moderately spaced, poorly to well coursed and of Chalk beach environment cobbles, with about 0.3 nipples per broken flint face (for a count of 20). Apart from rare pieces of iron slag the surface appears to be devoid of exotic boulders.

Palmer's Tower (Swinden's first tower, 1772) is again 'D' shaped (Figures 3.65 and 3.66). Turner (1970, 141) indicates that it is 7.62m. high. A modern wall of Scandinavian ballast boulders projects towards the south for about 5m. from the tower. To the west a short stretch of wall extending towards the position of South Gate is extensively repaired. The flint external face of the tower, which has a few putlock holes, is like wall '*M2*' to the east, patchy with repairs. Constructed of flints of a similar origin, the older portions appear to be those that are of broken to squared, moderately to tightly spaced, and well coursed. The nipple count here rises to as many as 1.0 per flint. Exotic boulders do not appear to be present. Apart from modern bricks, brickwork in the tower seems to be largely of Types 6/7 and entirely of 16th century date. In 1633, permission was granted for a windmill to be installed on the tower (*Great Yarmouth Town Wall*, 1971, 6). No evidence of this structure can now be observed.

The foundations of the once imposing South Gate (Figure 3.67) were according to Swinden (1772, 92) raised in 1344-1345. Turner (1970, 140), cited this date at possibly 1337-1338. She also drew attention to the print of the gate supposedly provided in *The Builder* (1886, **4**, 358) which she concluded illustrated a style that was 'clearly of mid-fourteenth century date'. Neither of the figures depicted in *The Builder*, which appear without text or caption (see Figures 5.1 and 5.2), however, are of the South Gate. The two flanking round towers, internally 9ft. (2.74m.) in diameter with walls 3ft. 9in. (1.14m.) thick, and the gatehouse between them, were pulled down in 1812 (Palmer, 1864a, 111). Swinden described the gate as square (Swinden, 1772, 97). To the west of the South Gate, as far as the River Yare, Palmer (1864a, 110) described the upper part of the wall, of which there is today no evidence, as being built more recently and of more slovenly appearance.

Swinden (1772, 96) describes the building of the South Mount in 1590 to the west of the South gate. A mound of earth 'much higher than the town wall, to command the river and the Denes adjoining, whereon were placed several large pieces of ordnance; the charging of raising it amounted to one hundred and twenty five pounds' (approximately £16,250 in 2006 terms). Vestiges of this Mount were said to remain in 1966 (unpublished Borough archivist's report) although nothing now appears evident.

CHAPTER 4

THE WALL FABRIC – AN ANALYSIS

4.1 Introduction

Great Yarmouth is one of comparatively few English settlements in which a substantial part of its medieval town wall is still standing. Traditionally, no doubt supported by the records of the collection of murage and the purchases made for materials in the period (Chapters 6 and 7), the wall is recorded as being built mainly during the first half of the 14th century (sections *1.4.3* and *1.4.4*). This analysis indicates that such a view is far too simplistic. Rather, the wall has been reconstructed on a variety of occasions so that visible workmanship relating to the 14th century has very nearly been obliterated by later activity. Despite the very limited range of building materials, essentially bricks, flints and a wide variety of exotic rocks, this detailed examination of the walls provides a much more extensive, if in instances tentative, history of the building and rebuilding of the town's defensive system.

4. 2 The Bricks

Seven major early brick types have been identified in the walls. Recognition of the individual types has principally been achieved on the basis of their size, method of manufacture and texture. The imported bricks (Types 4 and 5) were initially distinguished by means of their much paler, generally yellow, colour. Light coloured clay, however, also appears to have been locally and patchily present, for some Type 3 bricks can be similarly coloured. With the walls having undergone numerous repairs and alterations in recent centuries, more 'modern' brick types are common on the wall surfaces. For this reason, where the wall core has been exposed, the bricks seen far more readily assist in the identification of early episodes of building. The areas of uncovered core, however, in almost every instance reveal only fragmentary bricks. The widespread use of such bricks adds to the complexity of their identification, but tends to immediately infer rebuilding. Occasionally brick fragments are preserved in the wall with earlier mortar still adhering to a surface. The inference that the wall has been rebuilt because of the prevalence of reused brick material in its core could of course not necessarily be correct. The reused bricks could have been obtained from sources other than an earlier town wall. This must surely be the case with those brick fragments identified as Type 1. With just a couple of fragments observed at each of North West Tower (stretch '*A*') and King Henry's Tower (stretch '*B3*'), and a small number, with no whole bricks, in the Fishers Court area (stretch '*D1*'), this unusual brick type noted also by the author (together with P. Minter) elsewhere in East Anglia (Minter *et al.*, in press) is thought to have been made about the 11th century. Earlier buildings of this age are believed to have provided the source for these bricks.

Recently, Peterson (2007) made a case for a level of Roman occupation of the Great Yarmouth sand bank. It is an interesting reflection on Great Yarmouth's history that no Roman brick or tile fragments were observed in the walls. This surely supports the contention that during the Romano-British period the area of land which now forms Great Yarmouth did not support any permanent settlement. Behind the sand bar, a wide estuary probably occurred at the mouths of the Rivers Yare, Bure and Waveney, with Caister to the north and Burgh Castle to the south.

Of the seven principal medieval brick types determined, only four readily assist with providing evidence of the age of the walls: these are brick types, 2, 3, 6 and 7. The experience of both P. Minter and the author in examining the relatively uncommon imported bricks, Types 4 and 5, which were probably fabricated in The Netherlands but possibly, in some instances, in the Baltic Region, is such that they were reluctant to sub-divide them further. Type 5 was distinguished on the merits of only being observed at one locality (St Nicholas School, Dissenters' cemetery, section '*D1*'). Type 4 bricks probably range in size over time much as English brick types.

Tables 4.1 and 4.2 list the distributions of certain brick types and their respective quantities for different wall stretches. Table 4.1 applies to stretches of wall in which early bricks are exposed, especially in the wall core. Table 4.2 relates to similar analyses of the early walls of the towers. In Figure 4.1 the information from these tables is collated and the localities where these measurements could be made are indicated on a town plan. The distinctive characteristics of bricks of Types 2, 3, 6 and 7, in particular, and their more reliably determined periods of manufacture, enable wall stretches to be given a last approximate date of possible construction or repair. The wall stretches analysed fall into just two dates; either about 1450 or 1560. These dates seem to be associated with the end of the first building and repair phase, and the later building period when the walls suffered much rebuilding as rampiring was deemed necessary (see section **5.3**).

Exposed in recent years, to provide light to a window at St Nicholas School (stretch '*C2*'), large Type 2 bricks, of dates in the region of 1300 to 1350, were originally thought there to make up the sole composition of the wall core (Figure 4.2). Many of these bricks were whole. At the time it was advocated that this portion of the wall, was the only visible wholly unaltered area of the town wall built in the early 14th century. The discovery of a few Type 3 bricks in this exposed portion, however, indicates that even this wall was rebuilt, presumably about 1430-1450. Elsewhere in the wall, Type 2 bricks

CHAPTER 4 THE WALL FABRIC – AN ANALYSIS

Table 4.1 The distribution of the principal brick types observed primarily in the core elements of Great Yarmouth town wall.

Stretch of wall	Locality[1]	Overall % bricks cf. other materials[2]	Relative percentage of different brick types[3]				
			Type 1	Type 2	Type 3	Type 4	Types 6 and 7
'A'	West of Tower	25-35%	rare	rare	100%	-	-
'B2'	Town Wall Mews (inside)	5%	-	50%	-	-	50%
'B2'	Eden Place (inside)	uncommon	-	-	some	-	some
'C2'	St Nicholas School, wall gap	70%	-	100%	some	-	-
'D1'	Just north of Tower (outside)	60-70%	some	75%	3-15%	2-15%	-
'D2'	South of Tower (outside)	70%	-	90%	up to 5%	up to 5%	-
'E'	Market Gates walk (inside)	70%	-	c30%	c30%	some	c35%
'F'	Just north of Tower (outside)[4]	some	-	some	mainly	rare	-
'G2'	East Mount (external wall)[5]	60%	-	-	-	-	100%
'H'	South of East Mount (outside)	35%	-	10%	50%	10%	30%
'H'	Between Mount and Tower (outside)	50-60%	-	45%	10%	2%	43%
'J'	Opposite 'Time and Tide' (outside)	30-35%	-	70%	rare	some	30%
'J'	North of Garden Gate (outside)	40%	-	30%	c65%	some	-
'K'	North of Tower (outside)	30%	-	5%	10%	rare	85%
'L3'	North of Tower (inside)	15-25%	-	-[6]	rare	rare	c100%
'M2'	Between Towers (inside)	40-65%	-	rare	20%	rare	80%

Notes:

1. Approximate position. Core seen through wall section ('A' and 'C2' only), others from outside or inside of wall.
2. Other materials being principally flints and lime mortar. Insufficient bricks visible to determine percentage in two instances ('B2' Eden Place and 'F').
3. Figures often vary with slight changes in precise area reviewed.
4. The core is not properly visible in stretch 'F'. This brick distribution relates to the arrow slit construction.
5. This applies to the external 'brick' wall surface. Core not seen.
6. But in 'L3' to the north of area reviewed this increases to 20 per cent.

Table 4.2 The distribution of the principal brick types observed in the walls of certain towers in Great Yarmouth town wall. 'Modern' bricks have been excluded from these analyses.

Stretch of wall	Tower Locality and surface	Overall % bricks cf. other materials[1]	Relative percentage of different brick types[2]			
			Type 2	Type 3	Type 4	Types 6-7
'A'	North West (interior)	100%	5-10%	25-35%	-	60%
'A'	North West (exterior)	100%	-	-	-	100%
'B3'	King Henry's (interior)	20%	n/a	n/a	n/a	n/a[3]
'H'	Harris's (core)	45%	75%	20%	5%	rare

Notes:

1. Other materials being principally flints and lime mortar.
2. Figures may vary with changes in precise area reviewed. Additional replacement relatively modern bricks occur variously in each tower.
3. These bricks may be observed on the external surfaces.
n/a Not applicable. Access to the interior was not obtained, so that detailed analysis of the visible bricks was not possible.

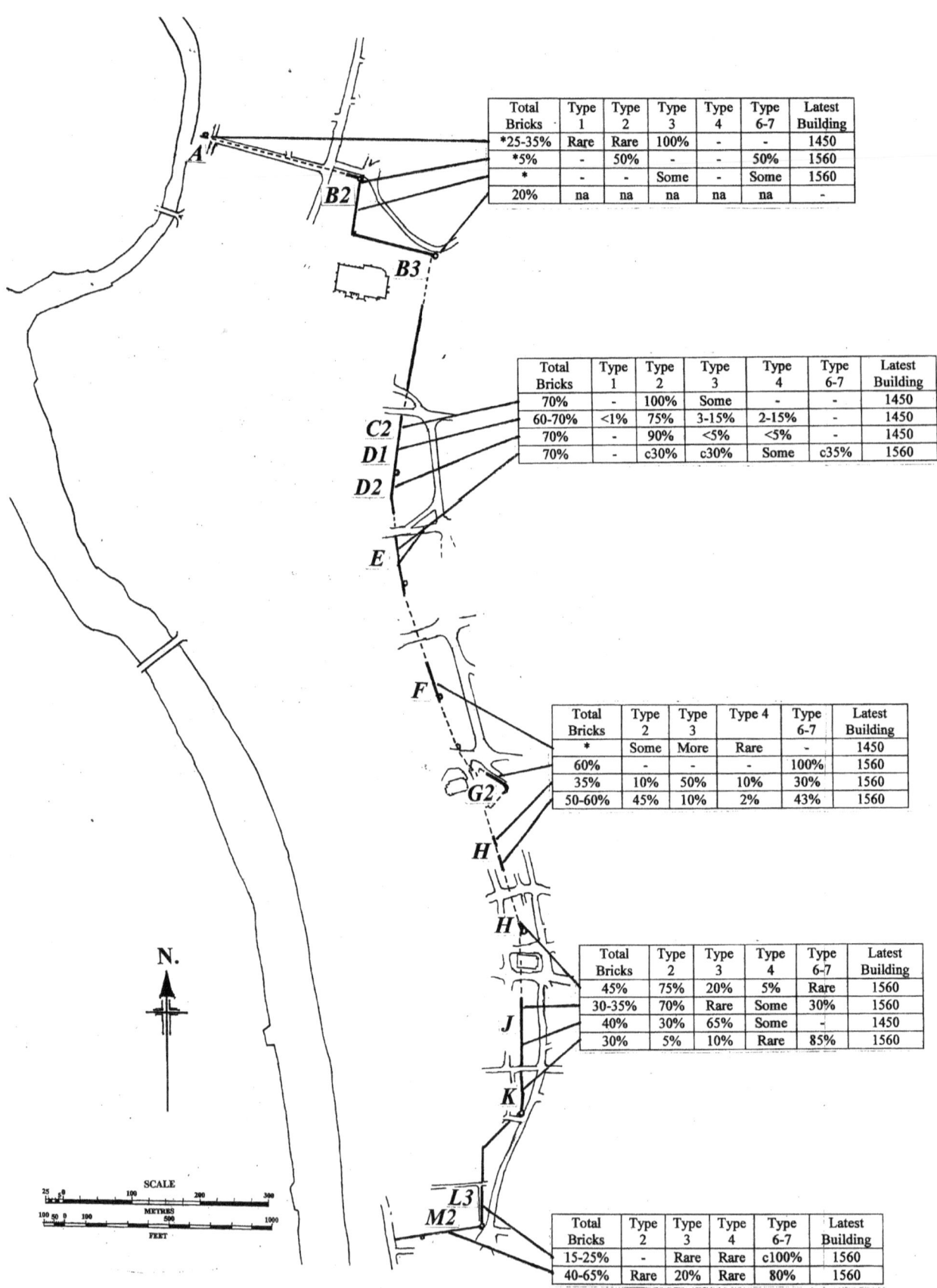

Figure 4.1 The distribution of principal brick types around the Great Yarmouth town wall, mostly related to the presence of wall core exposures (see also Tables 4.1 and 4.2). At locality 'B3' access and detailed measurements were not possible (na = not applicable). 'Modern' bricks have been excluded from these analyses.

CHAPTER 4 THE WALL FABRIC – AN ANALYSIS

Figure 4.2 The wall to the east of St Nicholas School (stretch '*C2*') has in relatively recent years been broken through to provide light for the classrooms. This view, from inside the wall and towards the north, illustrates the many Type 2 bricks visible in the wall core. One Type 3 thinner brick can also be seen. The flint wall face is on the right.

are generally fragmentary and they vary in their percentage of the total bricks, from being absent or rare to being as much as 75 (as in stretch '*D1*') or even 90 per cent (stretch '*D2*'). In areas where Type 2 bricks are so abundant certain features of the wall may be original.

Three other brick classifications remain; Types 3, 6 and 7. Type 3 bricks, mainly manufactured in the early to mid 15th century are present in almost all core exposures. Their quantity of the total bricks visible varies considerably, often over quite short wall stretches. Being relatively thin, in reuse they have a propensity to occur in a broken or fragmentary state. Instances occur, as in the inner wall firing bay archwork of the Market Gates shopping area (stretch '*E*') where they might possibly represent original, subsequently repaired, workmanship. Elsewhere, somewhat thicker, but similar, bricks and fragments occur in variable numbers: it is suggested that these bricks, most of which appear to be reused, may have been fabricated nearer to, or possibly within, the second half of the 15th century.

Type 6 and Type 7 bricks, all made in perhaps the first six or so decades of the 16th century, tend to merge in their characteristics. In some exposures (such as stretch '*H*') Type 6, manufactured in the early part of the century, predominate, perhaps to the exclusion of Type 7; elsewhere (as in the walls of the East Mount, stretch '*G2*'), the role is reversed. In certain of the visible wall stretches and particularly the brick portions of towers, Type 6/7 bricks outnumber all others and in certain parts of the wall (as in the East Mount walls and in parts of the wall to the north of Blackfriars' Tower, stretch '*L3*') they make up the entire or nearly the whole brick composition. In the East Mount walls all the bricks are intact, the documentary historical evidence as to the building date being fully substantiated by the make up of the wall fabric. In many places within the walls, rather than the towers, the Type 6/7 bricks are fragmentary in form. This may be observed, for instance, in stretch '*M2*', where many of the broken fragments are 'half' bricks. Whether such stretches of wall have been rebuilt or extensively repaired since the later part of the 16th century is difficult to determine. It seems very possible that they could have been built at a time of national emergency, perhaps in this instance the threat of the Spanish Armada, and would have used material from locally demolished domestic architecture. Domestic buildings constructed of brick would have been very much more common in the latter part of the 16th than in the first half of the 14th century. Furthermore, the mid to late 16th century, from the evidence of the historical records, was the period when the walls were rampired (section **5.3**). To provide the strength to retain these structures the walls would have had to be strengthened and areas of property close to the walls either buried or removed, providing an obvious source for wall material in the form of bricks.

The widespread reuse of bricks in the wall core presumably in part reflects good housekeeping, for it has to be remembered that the outside of the walls were improved in their appearance and quality with flint facework. The purchase of bricks for use in the walls (see section **7.2**) was recorded by Swinden (1772, 79-81, 84-85, and 88-92). About 500,000 bricks (*tegulae*) were purchased between 1336 and 1345 at a total cost of £52 10s. 11d. These were presumably Type 2 bricks which apparently cost on average 2s. 1d per 1,000 (at 2006 prices, approximately £55). With so much of the inner wall inaccessible for view it is difficult to assess how the half a million bricks were distributed or how near this was to the total brick supply for the early 14th century wall. No similar records have been found relating to what appears to be the second period of major wall rebuilding in the 16th century. As indicated above, this may result from a ready supply of bricks from immediately local demolished buildings.

Without detailed analysis of the brick compositions and textures it is difficult to determine the clay source of the different brick types. The texture of brick Type 2 suggests in this instance the use of river deposited clays and a local source seems probable. Bricks were made locally from river muds in the Cobholm Island area to the west of the River Yare in 1539 (Rutledge and Rickwood, 1970), and there is no reason why this should not have also been the case in the 14th century. The close proximity of Cobholm Island can be observed from its position at the southern margin of Faden's plan of Great Yarmouth of 1797, reproduced here as Figure 4.3.

The medieval town wall of Great Yarmouth, Norfolk, U.K.

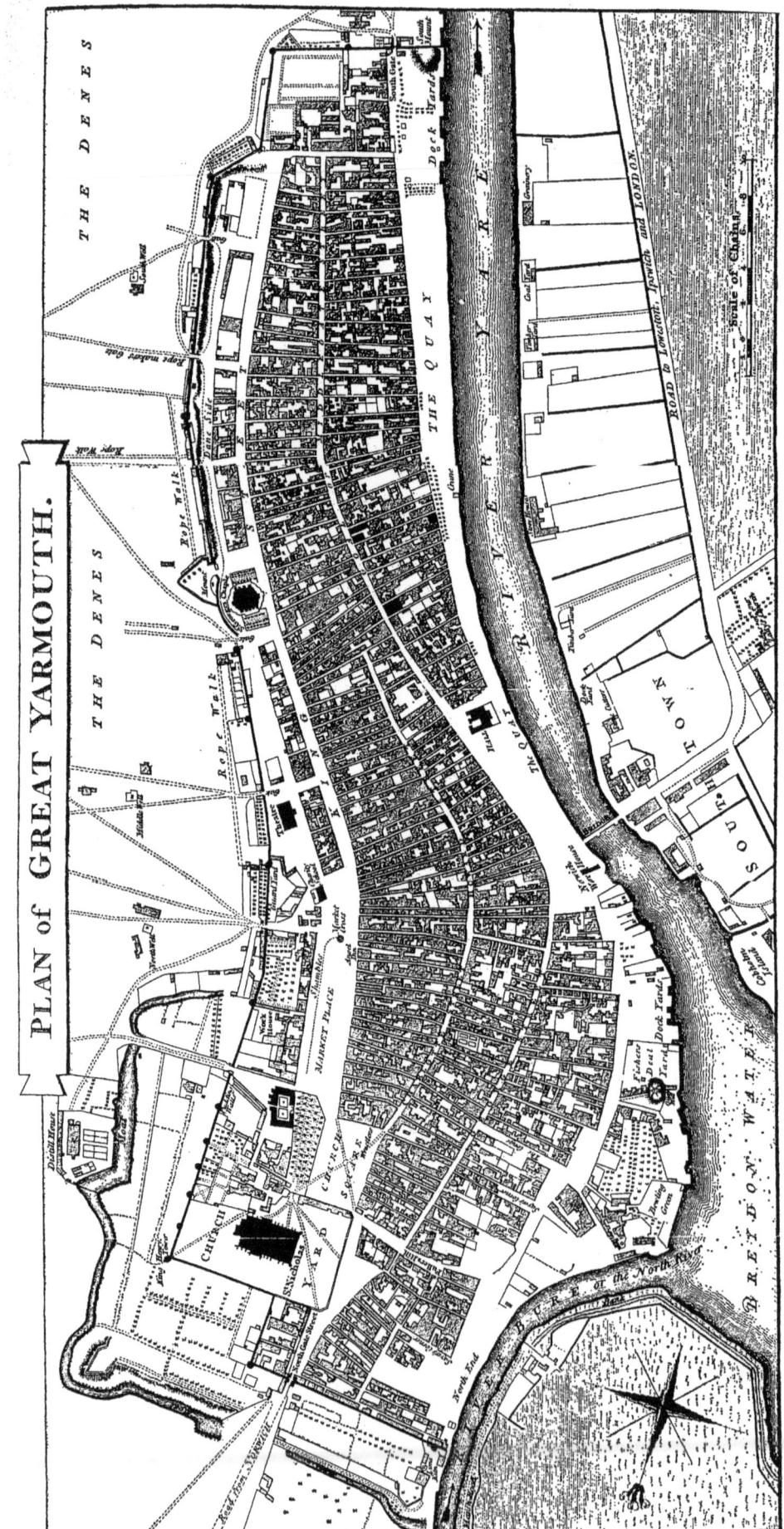

Figure 4.3 Faden's plan of Great Yarmouth dating from 1797; the map was published in *The Archaeological Journal*, 1980. This copy has been reproduced by kind permission of the Norfolk Records Office. Cobholm Island is marked at the western edge of the plan on the south side of Breydon Water. Note also the moat or ditch well outside the walls to the north of the town.

Table 4.3 The various rock types, here collectively described as 'exotics', which have been recorded almost in entirety on the external surfaces of the different stretches of the Great Yarmouth walls. The rocks fall into two main groups: those of igneous and metamorphic origin which have probably mainly arrived in the walls as ships' ballast from foreign parts and those of sedimentary origin. The sedimentary rocks make up a mixture of reused building stones (as Caen and Barnack Stones), glacially or fluvioglacially derived materials (mainly of Lower Cretaceous and Jurassic origin) and a smaller number of boulders of ballast.

Locality	Wall Section	Portion of Wall	Rocks of Sedimentary origin	Rocks of Igneous or Metamorphic origin
NW. Tower	A	Core Outside	Jurassic grey shale (R) Two yellow flagstone steps too high to identify (R)	- -
Town Wall Road	B1	Outside	Ssts; Jurassic 'featherbed'; Jurassic lst (with Rhynchonellids, Terebratulids); shelly Carboniferous Lst	Granite; gabbro; basalt
Ferrier Road	B2	Outside (N.)	Jurassic hard calcareous mdst; Chalk; sandy ironstone; crinoidal Carboniferous Lst (R)	Basalt; granite; gabbro; biotite hornblende granite; microgranite; quartzite
Ferrier Road	B2	Outside (S.)	Sandy ironstone	-
St Nicholas Churchyard	B3 (Bastion)	Quoins	Caen Stone	Granite/granodiorite (R), some later replaced by concrete
St Nicholas Churchyard	B3	Outside Core Inside	Caen Stone; Barnack Stone Chalk Orthoquartzites (some burnt); Green and brown ssts; Jurassic lsts and oolites; *Viviparus* lst; Caen Stone; numerous grave slabs (R)	- - Granite; microgranite; felsite; granodiorite; gabbro; basalt; gneiss; quartzite
King Henry's Tower	B3	Quoins	Caen Stone; replaced by Mid-Jurassic sst (?Ravenscar Group); later replaced by Mid-Jurassic oolitic lst	-
Priory Gardens	C1	Outside	Orthoquartzite; Jurassic ssts, lsts and oolites; Jurassic calcareous siltst; Lower Chalk	Granite; muscovite granite; diorites; tourmaline diorite; dolerite; basalt; weathered greenstone; gneiss; granitic gneiss; quartzites
St Nicholas School	C2	Outside	Ssts; Jurassic oolite; Lower Jurassic muddy sst (with Pectens and *Gryphaea*); Upper Greensand; Chalk; pink lst; iron slag	Granite; basalt; vein quartz; hornfels with feldspar porphyroblasts
St Nicholas School/Fishers Court	D1	Outside Inside	Calcareous mdst (R); Jurassic shelly lst (R); sst; London Clay with *Ditrupa* (R) Glauconitic micaceous sst (seats - R)	Granodiorite (R); granite and gabbro setts (R); diorite (R); dolerite (R); quartzite (R) -
Hospital Tower	D1-D2	Outside	Quartzitic sst; sst; shelly lst; saccaroidal decalcified lst; chert	Microdiorite; quartzite
Market Gates	E	Outside	Sst; Sandy lst; *Ditrupa* lst; lst; ironstone; Caen Stone	-

Location	Code	Side	Sedimentary	Igneous/Metamorphic
Pinnacle Tower (wall to north)	F	Inside	Sst; bryozoan rich lst	Basalt
		Outside	-	*Polydora*-bored basalt; quartzite
Shave's Tower	G1	Outside	Green sst; Yellow sst; Chalk	Microgranite; basalt
East Mount	G2	Outside	Caen Stone	-
Harris's Tower	H	Outside	Ssts. Chalk (in wall to north)	Granite; basalt
S. of St Spyridon	J	Inside	Quartzitic sst (R)	Tourmaline granite (R); basalt (R); vein quartz
'Time and Tide' Museum	J	Outside	Sst; Permo-Trias breccia	Granite; quartz-diorite; basalt; banded granodiorite; hornblendite; green quartzose schist; quartzite
		Core	Iron slag	-
		Inside	Quartzitic sst; ssts; Upper Jurassic grey mdst	Pale granite; granodiorite; basalt
SE. Tower (wall to north)	K	Outside	-	Basalt
SE Tower	L1	Outside	Caen Stone; *Viviparus* lst; Portland Stone (R); coarse micaceous sst	Tourmaline granite; microgranite; quartz diorite; basalt; quartz mica schist; pink quartzite
Blackfriars	L1	Outside	Caen Stone (some burnt); Portland Stone (R); Barnack Stone (some burnt); *Viviparus* lst; lst; Carboniferous? micaceous sst Chalk; iron slag	Microdiorite; gneiss; Pyrite-quartz vein
		Core Inside	Caen Stone (some burnt); Barnack Stone; *Viviparus* lst; Upper Jurassic calcareous mdst; sst; lsts; Chalk	- Granite; basalt
Blackfriars	L2	Outside	Caen Stone; Barnack Stone; ssts; lsts; Jurassic *Polydora*-bored muddy lst; Jurassic oolite with *Pentacrinites*; Chalk; iron slag	Basalt; gneiss
		Core Inside	Caen Stone (some burnt); Barnack Stone; *Viviparus* lst; Lower Jurassic shale; sst; Chalk	- Granite; granodiorite; basalt
Blackfriars	L3	Outside	Caen Stone; Barnack-like Stone; *Viviparus* lst; Mid-Jurassic oolite Chalk; iron slag	-
		Core Inside	Caen Stone; Barnack Stone; Lower & Upper Jurassic calcareous mdst; quartzitic sst; lst; sst; Mid-Jurassic oolite; Upper Chalk; iron slag	Granites; basalt; vein quartz; greenstone; gneiss; quartzitic gneiss; basalt; quartzite

Blackfriars Tower	M1	Outside	Iron slag	-
Mariners' Road	M2	Outside	Iron slag	-
Palmer's Tower	M3	Outside	-	-

Abbreviations:
(R) - Rock that is clearly a replacement (often set in different mortar)
sst - Sandstone
lst - Limestone
mdst - Mudstone
siltst - Siltstone

4. 3 Exotic rock types including reused material from earlier building

4.3.1 Reused rock material from earlier buildings

A distribution of all rock types other than flints built into the Great Yarmouth walls is illustrated by example in Table 4.3. The earlier buildings that were possibly the major contributors of stone, probably in the main ashlars, to the wall are listed in section *2.3.4*. There is a moderate quantity of two rock types only to be observed in reuse in the walls: Caen Stone and Barnack Stone, the former being the more abundant. Caen Stone, imported into England from Norman times, occurs as reused stone in wall stretches '*B3*' (outside and inside), '*E*' (outside), the East Mount '*G2*' (outside), and '*L*' (inside and outside). Barnack Stone, a harder limestone from Northamptonshire which was in use well prior to the Conquest, occurs only in stretches '*B3*' and '*L*'. Samples of both of these rock types are found exhibiting indications of having been burnt in stretch '*L*' (Figure 3.52). Both the stones and the historic evidence detailed for stretch '*L*' would suggest that this wall was rebuilt about the 1550s (see section *3.2.11*). Although the East Mount ('*G2*') was constructed about 1569 (Manship, 1619, 46), it was repaired and refaced in part with Caen Stone about 1588 (Manship, 1619, 47). The quality of the ashlars in the East Mount wall that was viewed is superior to that in the Blackfriars' wall. If it was refaced some 30 years after the '*L*' stretch wall was rebuilt, as suggested by Manship (1619, 47), this was undertaken probably with Caen Stone from a different immediate source, such as the St Nicholas charnel house and chantry (see Manship,1619, 40; and sections *2.3.4*d and *3.2.7*). The charnel house and chantry were shown as a ruin on the Cottonian plan of about 1585 (see Figure 5.3). The building or buildings which supplied the Caen Stone in the walls of stretches '*B3*' and '*E*' are more difficult to ascertain and the stone could possibly be from buildings much more recently demolished.

Pieces of *Viviparus* limestone (commonly described as 'Purbeck Marble' – see Potter, 2004), from the Lower Cretaceous succession of the Isle of Purbeck, occur within the walls of stretches '*B3*' (inside only) and '*L*' (Figure 4.4). The stone is normally polished and used for grave and tomb slabs or ornamental work. This use is often preserved in the shapes of the pieces found in the walls. Those found in stretch '*L*' presumably were initially used in the Blackfriars' monastery and church

Figure 4.4 The detail of a reused, weathered and broken slab of *Viviparus* limestone ('Purbeck Marble') observed in the flint facework of wall stretch '*L1*'. The numerous remains of the fresh-water gastropod *Viviparus* can be seen in the rock.

(see section *2.3.5*), those in wall '*B3*' may well be from relatively modern grave slabs.

Several small fragments of a flaggy micaceous sandstone of possible Carboniferous age (resembling York Stone), found in wall stretch '*L1*' might be interpreted as pieces of floor flags from the Blackfriars' monastery, their small size shed doubts, however, on this interpretation.

Much of the reused material from earlier buildings, in particular the flints, cannot be distinguished from the walls in which they are included. Bricks of different periods can be broadly dated and occasions of their reuse are more certain (see section **4.2** above).

In all the visible instances of wall core exposure the lime-rich mortar provides a very significant proportion of the core matrix. Instances occur where perhaps 40 per cent of the matrix is mortar. The provision of raw materials, such as lime and beach sand, for this mortar would have been considerable. Just as the wall bricks and flints were extensively reused, the present author believes, so was the mortar. Fragmentary reused bricks and flints in the wall possess little adhering earlier mortar, this having been removed before use. Powdered mortar from earlier periods of construction could have economically been remixed with additional lime for reuse at the rebuilding stages of a wall.

4.3.2 Exotic rock material

A wide range of rock types, here described as exotic, probably originated from the acquisition of ship's ballast. This must apply to virtually all the igneous and metamorphic rock types found in the external surfaces of the walls as rounded cobbles or boulders. Some of these originated from beaches, the *Polydora*-bored basalt seen in the wall near Pinnacle Tower being an obvious example. Although these rocks are too numerous and varied to have been examined in detail, the majority are unlike similar rocks from the British Isles and are thought to have come from the Scandinavian and Baltic regions. The second source of exotic rock types is principally of sedimentary origin, these rocks are less frequently rounded and more commonly irregular cobble size lumps. Many appear to be derived from areas of older rocks further to the north in Britain and they have probably been collected from soils overlying glacial till and water born derivatives of ice sheets (see also section *2.3.4*).

The distribution of these exotic rock types (Table 4.3), whether of ballast or formerly glacially transported origin, in the walls is of interest. Apart from those observed on the inside of walls at St Nicholas churchyard (stretch '*B3*') and the inside of the walls north of Blackfriars' Tower (stretch '*L*'), all others listed are found in exterior flint face wall surfaces. Numbers of rock varieties in the lists must reflect approximately an indication of the total abundance. It would be nigh impossible to argue that the surface of any large wall area was completely devoid of any exotic material. The presence of exotic rock materials within certain outer wall surfaces undoubtedly relates closely to the degree of access that has existed to those individual walls over the most recent centuries. Outer wall surfaces that are known to have had attached buildings, possibly since the 17th or 18th century (for example stretch '*M*'), are apparently without exotic cobbles. With, therefore, the exception of wall stretches '*B3*' and '*L*', there appears to be very little, or possibly no, evidence for the inclusion of exotic material in the walls as they were built. All the material appears to have been added to the exterior surface at a time of repair, probably the majority supplementing the surface in the last one hundred years. The inside wall on the north side of St Nicholas churchyard (stretch '*B3*') may well be similar, in its situation it has probably never supported attached buildings, and the presence of granite setts and broken grave slabs indicates the inclusion of relatively modern rock materials.

It seems probable that in the mid to late 16th century the inside of all Great Yarmouth walls became rampired. The rampiring appears to have covered the walls to the height of the sentry-walk (or nearly so). From that date, until such time that any rampiring was removed, the inside of the walls would have been impossible to see or repair and exotic material or reused ashlars could not have been added.

The detail of wall stretch '*L*' is different. Reused materials from the Blackfriars' monastery and church sites are used in abundance (see sections *3.2.11* and **4.2**). There is too, historic evidence of the use of ballast, for in 1551, the 'ballast ... of every fishing boat within town ... [was to] be laid ashore' (Ecclestone and Ecclestone, 1959, 48) in order to assist with repairs to a stretch of wall in the south which was described as 'decayed and falling down' (p. 47). This use of ballast must, in this instance, however, be critically considered. It is unlikely that local town boats would ride with ballast from as distant as Scandinavia; more probably local beaches would have provided a material such as flint cobbles for this purpose. Both the internal and external walls of stretch '*L*', with the exception of the internal wall of '*L3*' beside Tower Lane, were covered by attached buildings in the 1884 Ordnance Survey plan and much the same situation existed on the plan of 1928: by comparison with other Great Yarmouth town walls they should lack exotic cobbles and boulders. At least some of the included stone material, and certainly the reused ashlar stone, may well have been added in the 16th century. '*L3*' internal wall, bordered by Tower Lane, probably fittingly, displays more exotic rock types than any other wall in the town.

This is an appropriate place in which to discuss briefly any other details or records concerning the acquisition of ballast stones. Rutledge (1991, 11) noted that King John granted the lastage of the port (that is, the control of ballast) to a Henry de Hauvil, whose successors continued to enjoy the privilege until at least the late 14th century. This should have restricted the availability of ballast for use in the walls until this date. However, Swinden (1772, 79; and see *Great Yarmouth Town Wall*, 1971, 3) recorded the purchase of ballast (*lapid' de lastag'*) to be used as construction material. In 1336, 'a certain quantity of stones' was purchased from a 'Bartholomew de Thorp' for 2s. (in 2006 prices about £60). The quantity cannot have been very large for in 1345 another mariner was fined 6s. 8d. 'for casting his ballast into the haven'. The first citation appears to be the sole record of the purchase of ballast other than that described (above) in 1551, until recent records in more modern times. The ballast material of the 14th century has not been identified in the walls; it is most likely to have been either flint beach cobbles or early Flemish bricks carried on the return voyage following export to the Low Countries.

4. 4 Flints

4.4.1 Introduction

Upon first inspection the flint outer face of the walls of Great Yarmouth is indescribably complex. In particular, water penetration through the wall from the rampired rear of the wall must have assisted in the flint wall face frequently being removed and damaged over the years. Even the smallest stretches of wall, such as those in the towers, are made up of relatively small patches of varying flint types and styles of flint use. That parts of the flint facework are more recent than the wall core which they face is in some instances evident (see section *3.2.4*), but the variable pointing of individual flint face wall patches rarely assists in trying to determine their building sequences. Section *2.3.1* offers a brief generalised account of the recognised developing techniques in the selection and use of flints in building practice over time. Unfortunately, certain styles of flint wall face have been repeated with changing fashion, and the less complex and time-consuming styles of flint walling may date from any period of construction. Tightly packed squared flint face walls necessitating great skills in workmanship, for example, although probably first used in the early 15th century (as in Norwich Guildhall), returned to fashion in the Georgian and early Victorian periods. Detailed study of some of the walls (particularly stretches '*B2*', '*B3*', '*L1*' and '*L2*'. and an examination of Victorian and more recent flint walls elsewhere in Great Yarmouth), may have offered, however, certain clues as to the approximate sequence in flint walling styles in the Great Yarmouth defences.

4.4.2 Flint wall face styles

There are few, even moderately fixed, dates in the Great Yarmouth walls upon which the flint walling can be based. It can obviously be adjudged that nowhere can the flint facework pre-date the youngest bricks included in the wall core. Where they can be approximately dated, the inclusion of reused rock types, within the exterior walls may provide dates of further assistance. Walls '*L1*' and '*L2*' are believed to have been constructed in the 1550s (see section *4.3.1*), and their flint facework can be no earlier. It is believed that some flint wall styles were not first used before certain dates; galleting may extend back in time to the early 15th century, but is far more abundant in flint walls of about 1750. In walls '*L1*' and '*L2*' the galleting is presumably post the 1550s. Similarly, squared flints may have first occurred in walls in the early 15th century. The examination of Victorian and modern flintwork in Great Yarmouth revealed a specific style that was often visible in patches in the town's defensive walls. In that period, knapped, water-worn or quarried, cobbles of flint were typically placed in random and widely spaced fashion. More especially the broken surfaces lacked the presence of nipples (see section *2.3.1*iii). Whether nipples on broken flint faces were the result of using a fine pointed chisel in the past, the reaction of long-time exposure on incipient cracks (as proposed by Rose, 1860), or a mixture of both; their presence appears to assist in determining the possible relative age of the flintwork. Although the wall flints have apparently been much reused, the greater the frequency of the nipples, the older the wall face exhibiting them may possibly be perceived.

Carefully applying the information described in the previous paragraph provided a suggested relative time scale for various flint wall features:

5. Unbroken and unworked eroded beach cobbles of flint; often clearly used in recent patching and the tops of walls.
4. Work of typical 'Victorian' style as described in the paragraph above, no nipples unless flints reused and then number low to moderate but variable over area under examination.
This style can be variously modified; by, for instance, the inclusion of randomly placed bricks.
3. Areas of wall in which the flints are more tightly packed, either by the use of gallets and/or the flints being well knapped to squared: the work being generally, moderately well coursed (possibly all post 1550s).
2. Fairly widely spaced flints, knapped and generally in courses. Nipples are numerous.
1. Not identified.

If this classification is correct, work of the third category, with gallets and/or squared flints, occurs in parts of the walls of stretch '*L*' and might appropriately be dated to have been originally used in the wall in the mid 16th century. It would be necessary for aspects of the same style to continue until perhaps 'Georgian' times. The earlier style of use (category or style 2) is not particularly common but variously distributed in small patches. This is unlikely to be older than the oldest visible dating provided to date in this work by the bricks, that is, mid 15th century. If, as is believed, the earliest walling dates from early to mid 14th century, no external facework of this period appears to be present, hence the category or style 1 classification, 'not identified'. Because of the ability of flint to withstand weathering, it must be re-emphasised that flints first gathered for use in the very earliest wall facing, can be reused at any later period. This classification of wall face styles is also very general. For instance, style 5 flint cobble areas may well in some instances pre-date examples of adjoining style 4 workmanship. The classification of styles is summarised in Table 4.4.

Applying this classification to the wall flintwork that is much altered and patched opposite to the Time and Tide Museum (stretch '*J*'), it is possible to exemplify and relatively date the features that can be observed in this wall (see also section *3.2.9*). Figures 4.5 to 4.10 display wall face categories or styles in which all but the style seen in Figure 4.7 can be observed in this wall stretch. Style 2 is shown in Figure 4.5; two figures illustrate slight variations in style 3 (Figures 4.6 and 4.7); two varieties of style 4 may be observed in Figures 4.8 and 4.9, and finally style 5 is exhibited in Figure 4.10. Each of these figures is observed from much the same distance. The style of flintwork seen in Figure 4.7 has been selected from part of the wall of the North West Tower in order to demonstrate the complete range of flint patterns observed in the walls. An enlarged view of the different types of flintwork, each viewed from a distance of about 1.5m., for styles 2, 3 (two variants) and 4, each from stretch '*J*' are displayed in Figures 4.11 to 4.14 respectively. Each of these flint fabric styles, with the exception of that portrayed in Figure 4.7, is visible in a wall length of less than 100m.

From the excavation undertaken by Green (1970), it is clear that the flint face to the walls was applied from the earliest period of the wall building. The facework would

Table 4.4 Flintface styles identified in Great Yarmouth walls, and their typical age ranges.

Flintface Style	Typical features exhibited	Probable approximate age range
5	Unbroken and unworked, beach worn cobbles	1850 to Present Day
4	Knapped, water worn or quarried, randomly and widely spaced, cobbles. Nipples absent, except where flint reused, then small in number. Possibly ornamented with bricks, etc.	1800-1900
3	Flints tightly packed, either by use of galleting or well knapped to squared. Generally well coursed.	1550-1800
2	Fairly widely spaced flints, knapped and in courses. Nipples numerous.	1450-1550
1.	Not seen, but no doubt visible in wall face below ground level. Could be as style 2	1350-1450

Figure 4.5 Part of the external flintwork to the wall stretch 'J' (see section 3.2.8); this area being towards the northern end of the wall length near to Time and Tide Museum (approximately opposite No. 7, Blackfriars' Road). This style of flintwork appears to date from prior to the mid 16th century and is possibly the oldest style in the wall stretch (style 2). Notice the purposefully broken flints are well coursed and reasonably widely spaced (see also Figure 4.11).

Figure 4.6 A view of a different part of the same wall stretch 'J' as Figure 4.5 (approximately opposite No. 16, Blackfriars' Road). Here the flints are tightly packed but remain moderately well coursed. Packing, to present an almost continuous flint surface, is enhanced with flint galleting. This pattern is typical of style 3 and is thought to indicate a post 1550s age (see also Figure 4.12).

Figure 4.7 A variety of style 3 craftsmanship that is much less common than that seen in Figure 4.6 is here illustrated from the east face of the North West Tower. The flints are moderately well squared so that they fit closely and gallets are only rarely used.

Figure 4.8 Nipples are absent from the broken surfaces of these flints. This area of flintwork again occurs in the wall stretch 'J' opposite the Time and Tide Museum (approximately opposite No. 9, Blackfriars' Road). The workmanship is moderately coursed in this instance but elsewhere in the same area the flints are randomly related to each other, typical of Victorian work in the region of style 4 (see also Figure 4.14).

CHAPTER 4 THE WALL FABRIC – AN ANALYSIS

Figure 4.9 A further variety of Victorian (or more recent) flintwork is seen in this area of stretch 'J' wall (approximately opposite No. 8 Blackfriars'Road). In this instance 'half' bricks have been randomly added to the wall for supplementary ornamentation. The relationships and features of the broken flint cobbles are otherwise not unlike the style 4 (Figure 4.8) workmanship.

Figure 4.10 The battlements on the top of the stretch 'J' wall have been the most recently repaired. Unbroken cobbles of beach flints, well coursed and in this instance interspersed with mainly modern bricks, are typical of style 5 of the flint facework patterns.

Figure 4.11 Detail of style 2 flints in the area of Figure 4.5 of stretch 'J' of the Great Yarmouth town wall. Nipples, and opposing cups, are abundant on the broken exposed surfaces of the flints. Unbroken surfaces of many of the flint cobbles show curvature resulting from beach erosion. The lens cap has diameter of 50mm. This figure should be compared with style faces seen in Figures 4.12 and 4.13.

Figure 4.12 The detail of a style 3 flintwork example from the area of Figure 4.6 of wall stretch 'J'. Nipples and cups are fairly common, but the flints have been knapped to provide better fit and gallets built-in to fill any gaps. The lens cap is 50mm. in diameter, and the figure should be compared with style 2 work seen in Figure 4.11.

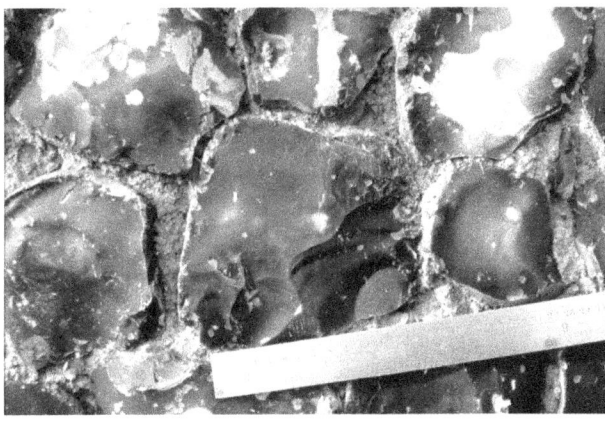

Figure 4.13 A further example of style 3 flintwork. This forms a very small patch in the area of Figure 4.6 and is more similar to the style seen in Figure 4.7. The lens cap is 50mm. in diameter.

Figure 4.14 The detail of the style 4 flintwork from the area of Figure 4.8 in wall stretch 'J'. Note the absence of nipples on the broken flint faces.

appear to continue perhaps three to four metres below the present ground level. Flint style 1 would presumably be evident in that part of the walls below ground level.

Some certainty can be expressed as to the type of occurrence from which the majority of the flints were originally collected. The mix of flints, in all but the most recent walls of style 5 (and some of category 4) above, and their proportions is indicative of collection from an Upper Chalk wave-cut platform. Such a coastline occurs in north-east Norfolk to the north of Cromer. Some of the flints appear to have been dug from the surface of such a platform whilst others were collected as not long dislodged cobbles. Those extracted from the surface exhibit both an irregular shape and elements of white cortex; the dislodged cobbles may show limited rounding and loss of much of their cortex.

4. 5 Summary

Although the historical record suggests that the standing Great Yarmouth defensive walls were constructed in the early to mid 14th century, no portion of this workmanship, even in the core of a wall, appears visible. This statement does not, of course, exclude the presence of work of this period which may remain present in the enormous stretches of the wall where the core remains unseen. Largely on the evidence of bricks of different appearances and ages, it is clear that the wall has been extensively rebuilt, often reusing materials, including bricks, which have been incorporated previously within the wall. A substantial amount of rebuilding or repair work appears to have taken place around the mid 15th century (Figure 4.1). A further period of particularly extensive rebuilding occurred during the mid to late 16th century, when rock materials such as those from the Blackfriars' church and monastery were used to supplement the building resources. The wall towers were it seems generally rebuilt during this period. This rebuilding period is apparently largely related to the necessity to strengthen the walls prior to embanking the wall with earth (rampiring).

The outer face to the walls, which is mainly of flintwork, has very frequently been independently repaired or replaced. Although a speculative sequence of different patchwork styles for the flintwork has been proposed (Table 4.4), no style has been identified that would appear to be datable to the earliest 14th century wall. It should be visible in the wall faces obscured below ground level. The widespread patching of the outer wall surfaces makes their relative dating and interpretation extremely difficult. Furthermore, the enormity of the extent of the patching through the ages makes it relatively impossible to precisely detail the detail of the work without a stone by stone drawing of the faces of all the walls. Ballast and other exotic rock types included in the external, especially outer, wall surfaces mainly appear to have been introduced, in wall repairs, in recent centuries.

CHAPTER 5

OTHER DEFENSIVE SYSTEMS: EARLY WALLS, TOWERS, RAMPIRES, PSEUDO-RAVELINS AND A MOAT

5.1 Early walls

The historical evidence for the initial and hesitant erection of the currently visible town walls has been presented in section *1.4.3*. Commenced in earnest towards the beginning of the 14th century the wall may not have been completed until as late as about 1396, approximately 80 years later. The earliest defensive systems in most medieval English towns consisted of banks and ditches frequently erected in Norman times, their replacement, ultimately by stone walls, commonly occurred as much as a hundred years before the 14th century. There is ample evidence that Great Yarmouth supported a castle and though the date of its construction is unknown it was probably built before 1208 (Rutledge, 1990, 43), it became a gaol in 1550, and was pulled down in 1621 (see sections *1.4.1; 2.3.*3a and *2.3.3*e). Typically, castles are further protected by a system of defensive walls and Great Yarmouth would be expected to have possessed such walls, perhaps initially of wood and later of stone. The evidence, however, is very limited. It appears to be almost confined to a statement by Swinden (1772, 80) concerned with the costs of demolishing an old wall outside the town in 1336 and 1337, which inclusive of tools and transport (to the wall being built) amounted to £6 2s. 11d. (in 2006 terms, about £3,000). No indication of the types of stones brought to the 1336/7 wall from this source has been observed but if they were flints they would not be identifiable. Inflation during the 1330-1340 decade was high but it is difficult to explain the outlay when it is compared (also from Swinden, p. 87) with a single boat load of flints ' bought and received of a certain mariner of Cromer' in the same period at a cost of 5s. (about £125 in 2006). An old wall may, of course, need not have been constructed of flints, nor been part of a prior defensive enclosure, but it would appear to have supplied a considerable amount of material. It could have been also walling funded and built in 1285 (see section *1.4.3*), erected under the provision of the earliest recorded financial support.

Further references were also made to town walls specifically in the Blackfriars' area in Page (1906, 436). In 1287, 'Much of the town walls were destroyed, and the house of the Dominicans was covered by the waves'. In 1290, a scheme to move the town wall in that area was abandoned (*Inquisitions post mortem* 18, Edw. I, No. 140). A plan of the town produced by John Deleny in 1734, held in the Norfolk Records Office, interestingly indicates 'old fortifications' on the denes some distance to the south-east of the Blackfriars' area.

The shape of the 14th century walled enclosure to the town, or any previous defensive enclosure would have been determined by a number of important buildings known to exist in the town at the time. These buildings must have included the castle (close to where the church {theatre} of St George now stands, Ecclestone and Ecclestone, 1959, 84), the large church (according to Manship, 1619, 36, with as many as 6,000 communicants) and associated Benedictine priory of St Nicholas completed in 1119 (Pevsner and Wilson, 1997, 494), and Blackfriars' Church and monastery, as well as other buildings. Because the castle and the St Nicholas and Blackfriars' sites were established more than a hundred years before the completion of the 14th century wall, both the present town wall and any earlier wall are likely to have followed the same course in skirting the properties. If it existed, the line of any pre-14th century wall cannot otherwise be currently deduced for certain (Pevsner and Wilson, 1997, 517).

5. 2 Towers

The towers to the town wall circuit must have originally provided the strongest elements of the defensive system. With the onset of potential attack from cannon fire in the 16th century their walls would have been more vulnerable. All appear to have been reconstructed during the period of improving the state of the town walls about this time. Originally, it appears likely that the towers had a flat roof, for Palmer (1854, 275) and others comment that lead was purchased 'for the cover of the towers'. According to Palmer (1854, 276) many had pieces of 'large ordnance' placed 'on them'. Each of the towers that remain exhibits elements of alteration and 16th century bricks predominate in their structure. Unfortunately, this examination was largely restricted to the lowest readily visible, external walls of the standing towers, and the construction materials at both higher levels and internally were not examined.

Entry to a few of the towers proved possible and these appeared to confirm the statement by Turner (1970, 63) that 'at Yarmouth there is no trace of an internal stair' in the towers. The lowest chamber could always be accessed from inside the wall at ground level. The floors above originally were entered from the first floor wall walk. The intention of this arrangement was presumably to prevent any external aggressors who might have gained access to the sentry-walk easily achieving entry to the interior ground level. Access to the sentry-walk from the interior ground level may well have been by the use of removable ladders, although internal stairs may have existed in the no longer standing gate towers.

Figure 5.1 An engraving of the South East Tower as it appeared in *The Builder*, 1886. The tower is viewed from the south-east and clearly displays its chequerwork.

Figure 5.2 An engraving of Blackfriars Tower reproduced from *The Builder*, 1886. Both this engraving and that in Figure 5.1 were completed by Mr G. Ashburner and they appeared on page 357. The tower is viewed from the east.

It has been suggested that tax remission in 1512, provided for obtaining cannon and 'making defence' (*Calendar of State Papers*, 1509-1514, 687) may have been used to produce the brick and flint chequerwork set within square panels, which is such an attractive feature of the South East and Blackfriars' Towers (*Great Yarmouth Town Wall*, 1971, 4)(Figures 5.1 and 5.2). Much of this chequerwork, however, shows indications of more recent repair. Further tax and fee-farm remissions were granted until the 1550s (*Letters and Papers of Henry VIII*, fo. 129). Both chequerwork and the unusual conical tops to some of the towers have been attributed to the same period. They are featured, for instance, on the Cottonian map of about 1585 (Figure 5.3). It should, however, be pointed out that the picture-map of 1588 (Frontispiece, and see Figure 5.6) and other plans fail to show such conical roof structures. Neither a conical roof nor chequerwork contributes to the defensive qualities of the towers even if these features might have impressed a potential aggressor.

5.3 Rampires

In 1545 Henry VIII, then at war with France, instructed the Duke of Norfolk to inspect the state of Great Yarmouth's fortifications. The techniques of battle had altered from the time of the 14th century wall's first construction. Sword and arrow had become less effective than cannon and artillery. Although the cannon had been invented in the early 14th century, it was not until the 15th century when cast barrels of some sophistication appeared that gunpowder for propulsion of projectiles began to replace the catapult principle in artillery. The pistol had been introduced to the English cavalry a year earlier (Carman, 1955, 132). The Duke's review reported on the unsatisfactory state of the walls (Palmer, 1854, 417). The simplest manner by which walls at the time could be strengthened was to embank them internally; this not only offered greater protection from cannon fire but permitted defensive artillery to be raised to a level where it could be fired from inside the top of the wall.

According to Carter (1980b, 303) internal wall embanking in Great Yarmouth 'was not carried out until after the 1539 inspection of the east coast defences'. Manship (1619, 73) indicated that the wall between Market Gate and Blackfriars was rampired between 1544 and 1587. Still further rampiring occurred in 1588, particularly over the same Blackfriars' walls and near the St Nicholas Priory, presumably at this time imposed by the threat of the Spanish Armada (Palmer, 1864a, 107-108; Rutledge, 1963, 127; citing Manship, 1619, 73-74).

CHAPTER 5 OTHER DEFENSIVE SYSTEMS

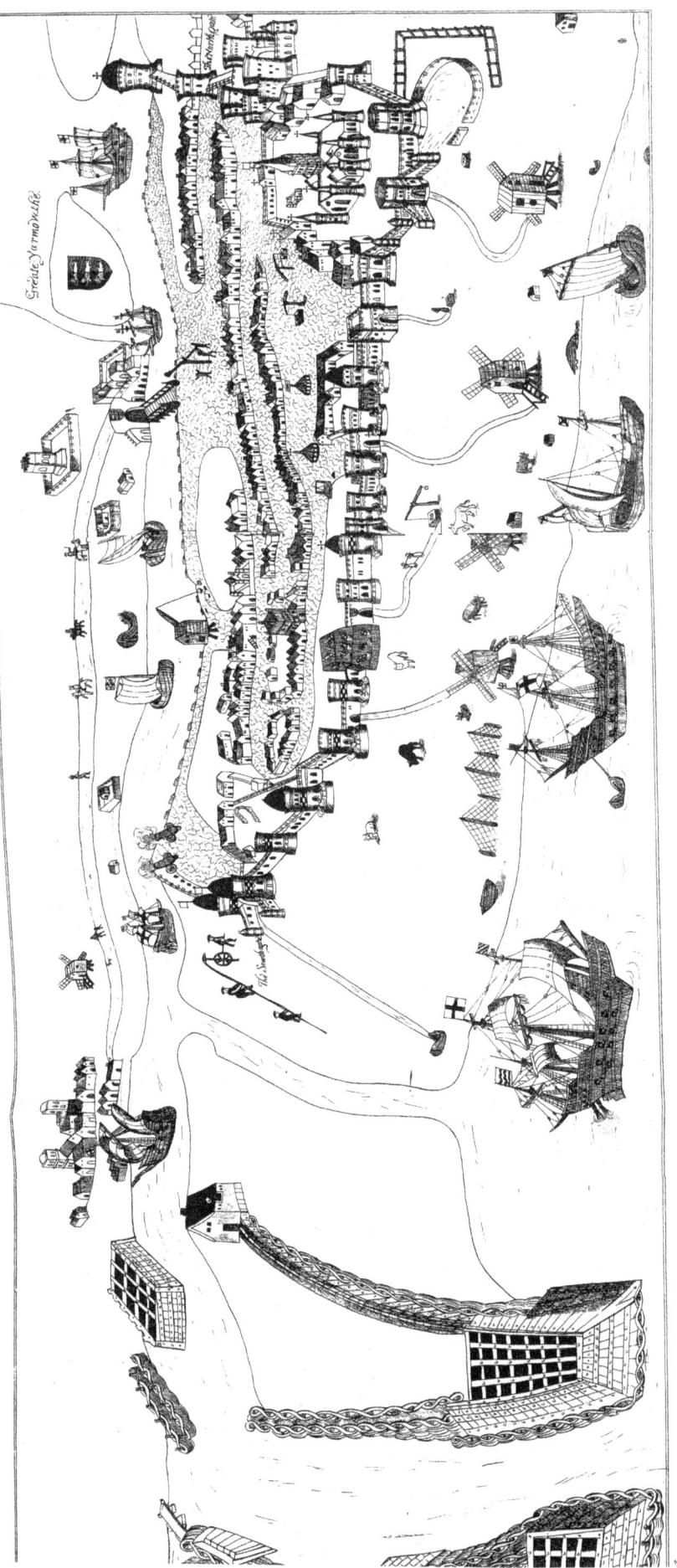

Figure 5.3 A copy of the picture-map of Great Yarmouth reproduced by Palmer (1854). Geographical north is to the right. The original copy of the plan which is thought to date from about 1585, is often referred to as the Cottonian map. It is held in the British Library. Reproduced by kind permission of Norfolk Records Office, Norwich.

Figure 5.4 The rampire to the north side of the graveyard at St Nicholas Church viewed from the west. All possible firing bays on the inside of the town wall have been buried beneath the earth embankment which rises to the wall from the south.

Dates within this range of 1544 to 1588 are also cited by both O'Neil and Stephens (1942, 5) and Ecclestone and Ecclestone (1959, 49, 51, 54). In 1553, concern that the town might have to be realistically defended caused the gates into the town to be rampired (Ecclestone and Ecclestone (1959, 52, citing the *Great Yarmouth Assembly Book* of 12th of July of that year). Palmer (1864a, 108) more correctly advised that the North, South and Market Gates remained in use. Swinden (1772, 95) suggested that the rampires were raised to the top in 1587 and that they were then of the order of 12m. in breadth from the wall (Manship, 1619, 73). Swinden (1772, 93-96) also noted that in 1545 the 'Duke of Norfolk...caused all the little hills, or sand banks, without the town, on the Denes, which the sea and easterly winds had accumulated to be by the inhabitants conveyed into the town'. That this material was used to help to create the rampires is confirmed by Manship (1619, 73). He states that the sand was 'brought in by the townsmen; and by that means the whole town, within the space of fifteen weeks...was...strongly fortified'.

At the present time the best example of rampiring may be seen in stretch '*B2*', the west-east wall on the north side of St Nicholas churchyard, where the rampire rises well up the inside wall (Figure 5.4). Most of the town walls show some evidence of this embanking (Figure 5.5) although often it is represented simply by much higher land existing on the inside of the wall. It is clear that in rampiring the walls the earlier firing bays would of necessity be covered by the earth embankment and no longer available for use. In every instance visible the arched bays are repaired or part reconstructed in bricks which were dated as of Type 6 (early 16th century) or Type 7 (1530-1560) age. This would indicate that immediately prior to rampiring the walls had to be rebuilt and strengthened, an appropriate precaution in view of the potential thrust from the embanking to be placed against the wall. In certain instances it appears that it was decided to remove the firing bay piers and the existing sentry-walk altogether prior to building the rampire, to leave a flatter wall to embank against. This is particularly evident in the north-facing (as stretch '*B1*') and south-facing (as stretch '*M*') walls. The east-facing walls, facing the direction from which attack was normally expected, retained the piers. During the period within which the rampires were stated as being constructed, 1544 (possibly 1539)-1588, one wall stretch (within stretch '*L*') is described as having fallen down (see section *3.2.11*b). This may have occurred in 1557. There is no evidence for this collapse extensively affecting the more stable and strongly built firing bay piers, which still in part remain.

The rampiring of the walls would have had two further important impacts. The embankments extended 12m. from the inside of the wall. They would have caused any prior buildings or properties up to this distance from the wall to have been either demolished or buried. Notwithstanding the fact that the number of brick built properties must have been limited, the demolition may have provided some bricks for reuse in the wall. This could well explain the large percentage of 16th century bricks that appear in the repaired inner walls in a broken state; the broken bricks being especially of the early 16th century, Type 6 brick. Secondly, the firing bays would initially have been fronted with crossbow loopholes or arrow slits. Following rampiring, the arrow slits, often bricked up or highly modified, remain. They would have become inaccessible from inside the wall and a few may still remain in this state (as some in stretch '*F*'). Their modification to gun ports would have been expected if put to 16th century requirements. As it happens, just a small number, in stretch '*L1*', in the wall immediately to the south of the South East Tower, have been modified in this way. It would seem possible that following rampiring, access to these few firing bays was retained, either by means of steps down through the rampire or a

CHAPTER 5 OTHER DEFENSIVE SYSTEMS

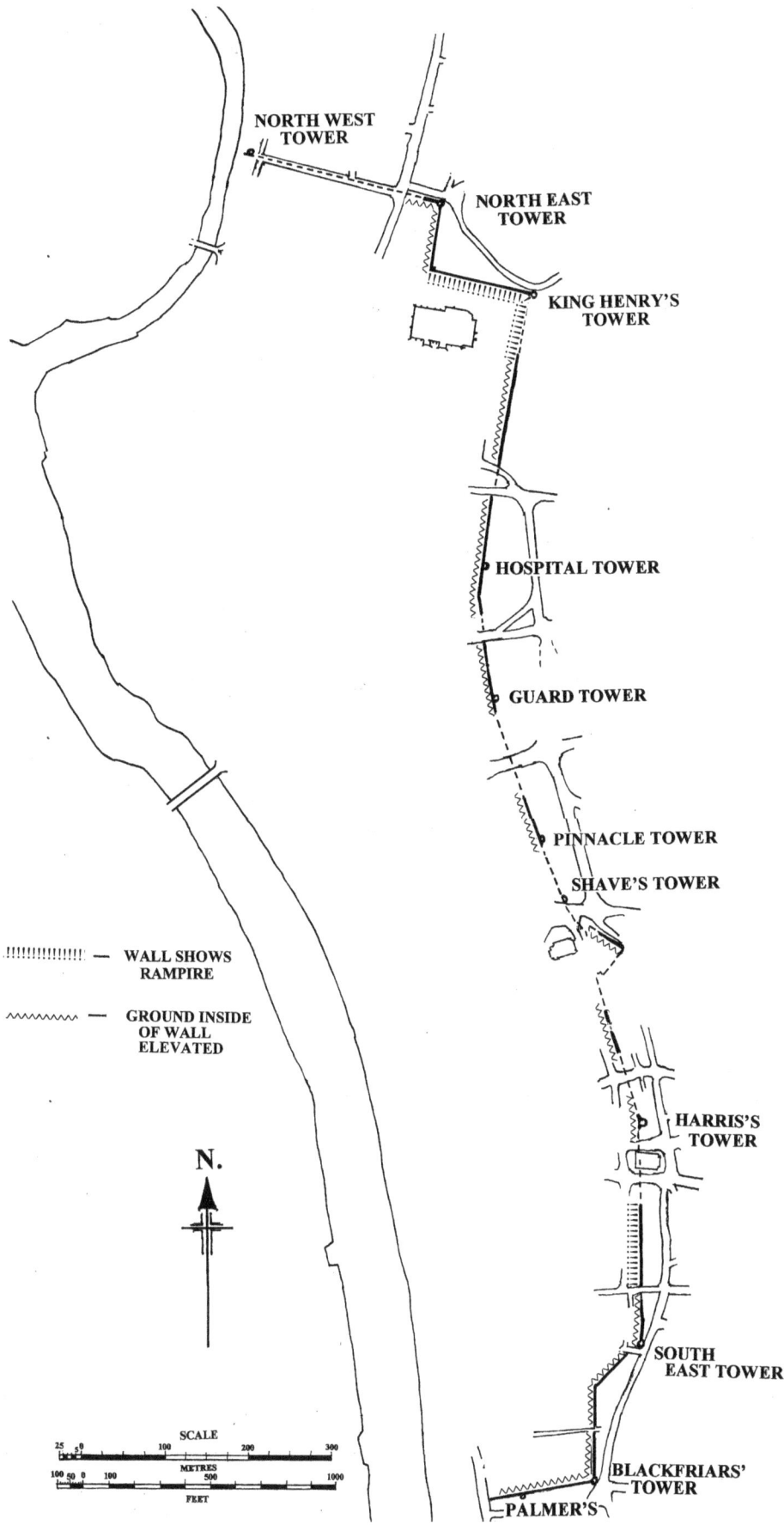

Figure 5.5 Areas of the Great Yarmouth town wall which indicate evidence or limited evidence of the presence of the 16th century rampiring.

short tunnel from the South East Tower. In this position they could have provided additional small arms fire cover for the tower.

Over much of the inside of the town walls the 16th century rampires have subsequently been removed or partially removed (Figure 5.5). There is limited written evidence of this removal. Palmer (1852, 391) does, however, refer to workmen 'lately employed in levelling the rampart' in the region of Blackfriars' Tower.

5. 4 Pseudo-ravelins or Mounts

A number of gun platforms or earth mounts were constructed around the periphery of the Great Yarmouth walls during the second half of the 16th century and the early years of the 17th century. Being triangular in shape they have been referred to as ravelins (Manship, 1619, 74; and others). More recently, Pevsner and Wilson (1997, 519), for instance, refer to the only mount that still in part stands, as a portion of a unique English survival of a ravelin. The structure is more appropriately described

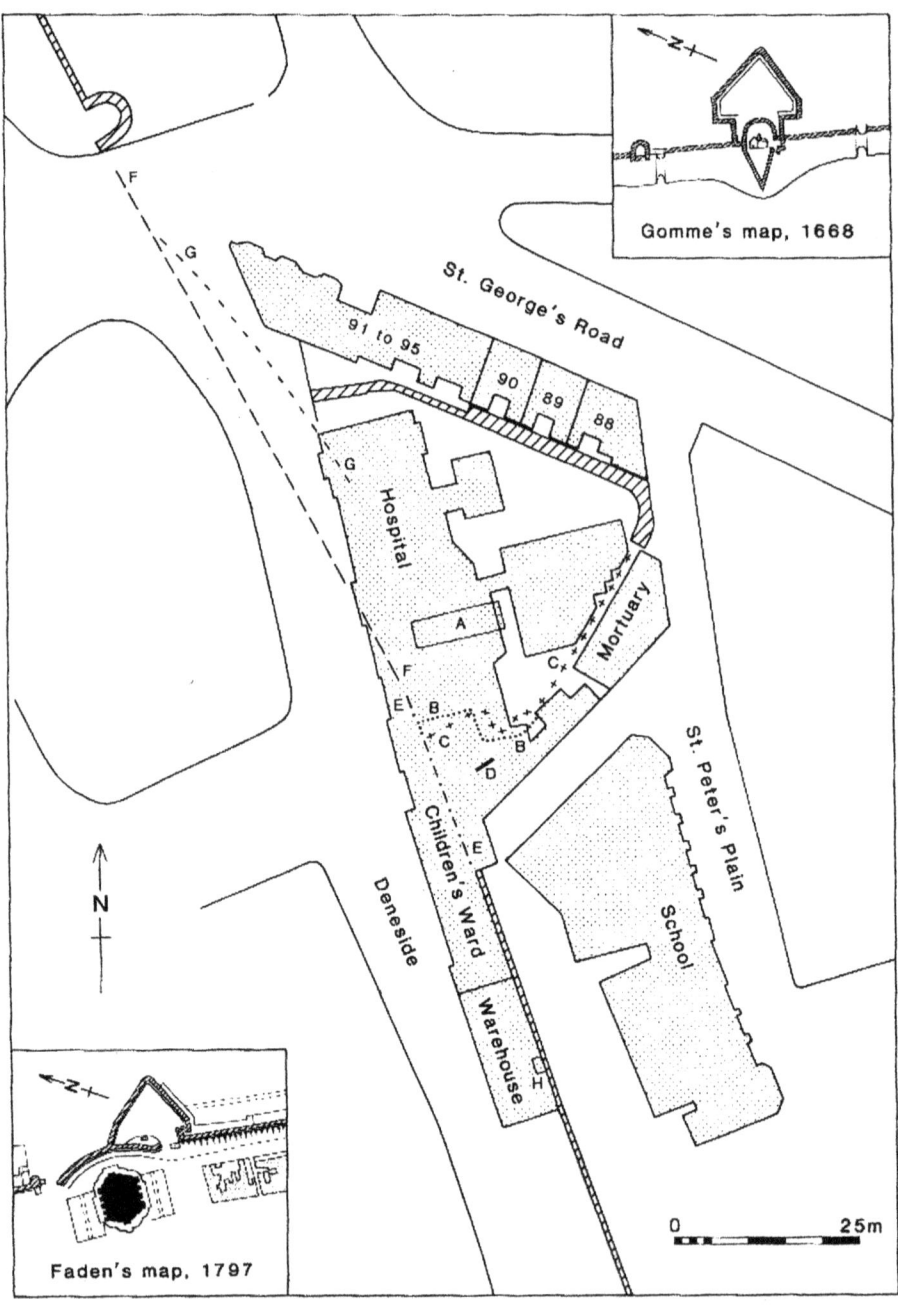

Figure 5.6 A plan of the East Mount, Great Yarmouth created by Rose (1991) in 1984 at the time of building demolitions and excavations at the site. The plan is reproduced by the kind permission of the author and the Norfolk and Norwich Archaeological Society. Of the inset plans, Faden, 1797, is reproduced as Figure 4.3 within the present work. Gomme's 1688 plan is not included in this work. Figure 3.30 was taken to the rear of Nos 89 and 88, St. George's Road towards the south-east.

Chapter 5 Other Defensive Systems

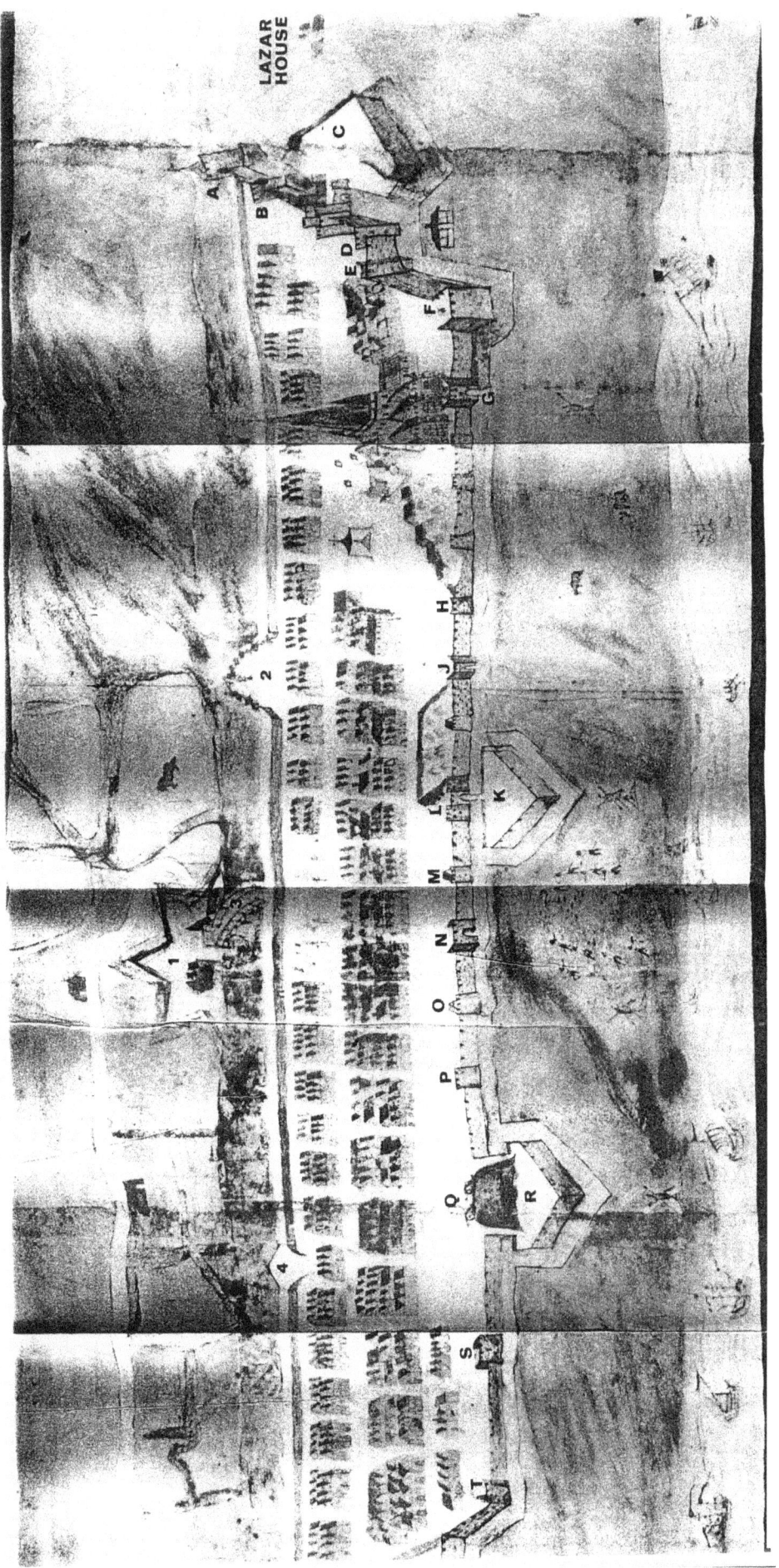

Figure 5.7 A more greatly enlarged example of the Hatfield House picture-map which appears as the frontispiece to the present work. The plan, here displaying most of the town walls (north to the right), is thought to date from about 1588 and have been prepared in preparation for the possible invasion of the Spanish Armada. The plan depicts the East Mount (K) as a true ravelin, although there is no evidence that it was ever built in this form. Note also the absence of any moat external to the walls and King Henry's Tower (F) being illustrated as square. The plan is reproduced by kind permission of the Norfolk Record Office.

The Medieval Town Wall of Great Yarmouth, Norfolk, U.K.

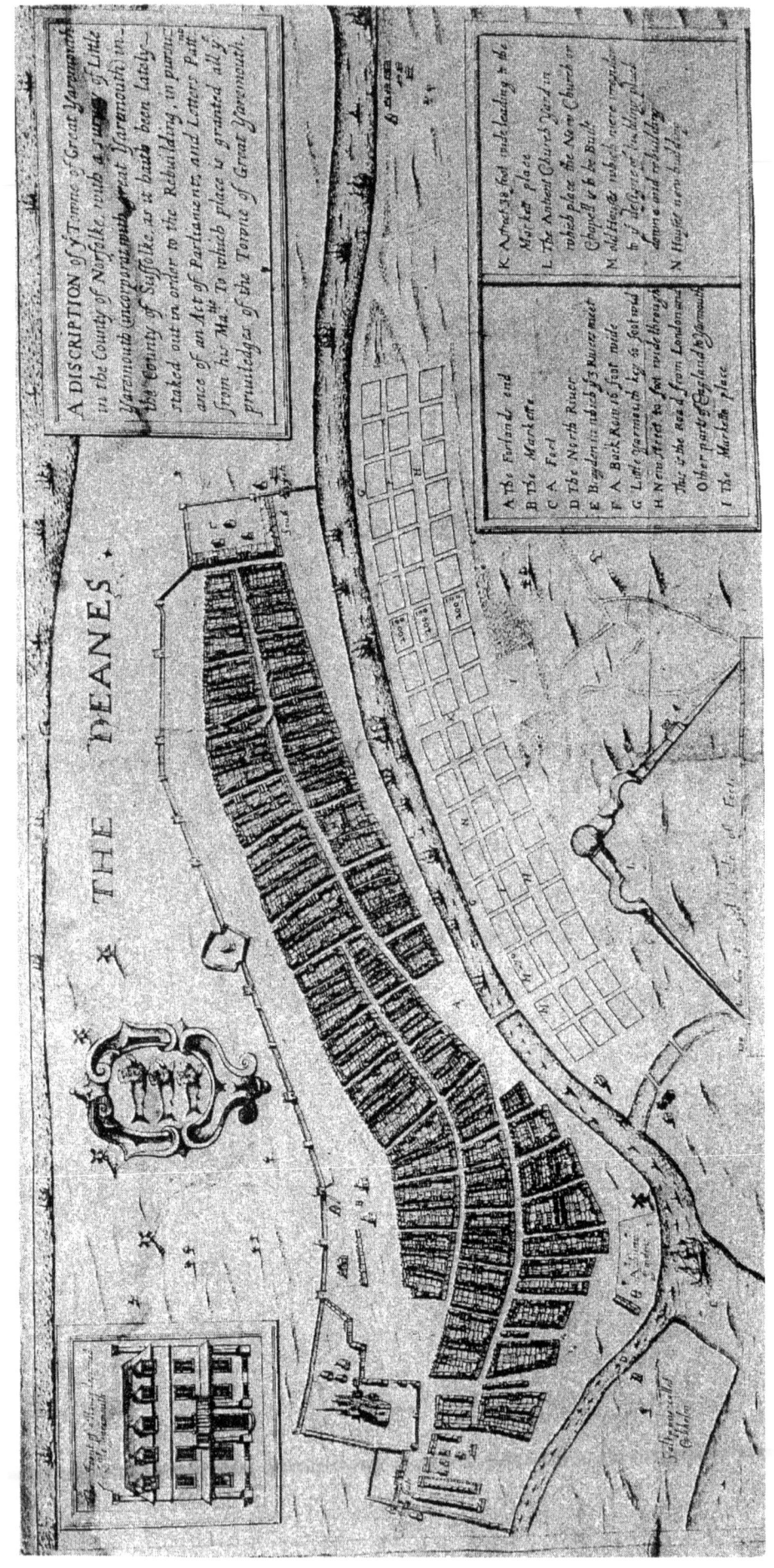

Figure 5.8 The plan of Great Yarmouth prepared by Sir Robert Paston in 1668. Geographical north is to the left. This copy was reproduced from Rutledge and Rutledge (1978). The plan is held in the Norfolk and Norwich Archaeological Society Library. Permission to publish is gratefully acknowledged.

using the title to which it is normally referred; as the East Mount (see section *3.2.7*). Manship (1619, 46-48) described the actions taken by the citizens of Great Yarmouth in 1588 to protect the town from the Spanish Armada, these referred only to the building of the East Mount, a smaller South Mount, and a boom to protect the harbour. He indicated that the cost of these items and their construction amounted to £1,165 8s. 4d. (in 2006 terms about £163,000) plus labour.

Whilst the Elizabethan picture map reproduced by O'Neil and Stephens (1942) indicates a considerable number of gun platforms presumably planned for a potential Spanish invasion in 1588 (see frontispiece and Figure 5.6), rather less appear to have been built. Those referred to by others are: the New (later Chapel) Mount initially built in 1569-1570 (Swinden, 1772, 96-98; Palmer, 1854, 276), this being the sole earth mount to remain and now called the East Mount; another mount or 'raveline' constructed about 1588 in the vicinity of Blackfriar's Tower (Palmer, 1864a, 108), the precise whereabouts of which were uncertain by 1864; the South Mount built in 1588 (Manship, 1619, 47) or 1590 (Swinden, 1772, 96; Ecclestone and Ecclestone, 1959, 86); 'Symonds' Seat Mount, just north of Garden Gate (Palmer, 1864a, 114); and the Main Guard (later the Battery Mount, but see below) with encompassing wall higher than the town wall, which extended south from Market Gate, in 1626 (Swinden, 1772, 99). Although listed as gun platforms, some of these structures were in fact elaborate rampires and certainly bore no resemblance to ravelins. Emery (1998, 1) referring to the Main Guard, for instance, describes it as, 'a substantial rampart which was thrown up along the back of the town wall between 1569 and 1650, and which stretched from Market Gate southwards as far as a position on the wall to the rear of… Guard Tower'. This description would appear to relate to the rampire constructed in the same position. However, just outside the wall and immediately south of Market Gate, Rye (1972) described an excavated area (at TG 525 076) of possible early 16th century bricks. From their description these bricks resembled Type 6, but they were set on edge and heavily mortared. He postulated that the area represented part of a gun platform or earth mount and possibly this might correspond to a portion of the Battery Mount.

In 1991, Rose published an account of his studies of the East Mount covering the period 1984-1986, when extensive demolition and building alterations were being implemented in the area. A site plan from his article is reproduced by kind permission here as Figure 5.7. Figure 3.31 is of the wall face behind number 89, St George's Road, viewed towards 88. Rose was able to determine the true shape of the Mount (dotted line B-B) and establish that it was never a true ravelin. The line of the earlier town wall, E-E and F-F, was also ascertained, although the position of the early gate was not discovered.

It should be noted that the plan prepared by Paston in 1668 (reproduced here as Figure 5.8), shows the obvious presence of only a single, ravelin type, structure, that of the East Mount. Additionally, the South Mount is shown simply as an unwalled mound against the inside of the town wall in the plans of Swinden, probably of about 1758 (Figure 3.2) and Faden, 1797 (Figure 4.3). The Faden plan extends northwards of the town wall to also indicate a north-east and a north-west partially walled mound. These are placed inside a moat constructed well outside the town walls. They were presumably built about 1642 at the onset of the Civil War (see section **5.5**), when imminent attack appeared possible by land from the north.

5. 5 A Moat

5.5.1 An early 14th century structure

It was not uncommon for medieval English walled towns to also be defended by a moat. The evidence in Great Yarmouth's case for the whereabouts, and possibly the existence, of such a structure currently is tenuous. The point should be made that Great Yarmouth town was, and still to a less extent is, largely surrounded and bordered by water, which immediately suggests that the requirement for a moat was unnecessary. There are certainly limited references to the construction of ditches at dates prior to the 14th century. In 1261, Henry III granted by letters patent licence to enclose the town with 'a wall and ditch' (*Calendar of Patent Rolls*, 1258-1266, 177), at the same time a murage grant was provided to assist the work involved. An element of this work had been undertaken by 1285 (*Calendar of Patent Rolls*, 1279-1288, 328), although in the intervening period the monies received had to be returned (*Calendar of Patent Rolls*, 1272-1281, 315). Quite naturally, the assumption has been that any ditch dug would have been parallel to, and adjoining, the wall. Nothing survives above ground of the ditch, but its construction might have been underway in 1308 when complaint was made that the town had begun to raise a dyke to the prejudice of neighbouring Little Yarmouth (*Calendar of Patent* Rolls, 1307-1313, 124; Great *Yarmouth Town Wall*, 1971, 2). Rutledge (1999, 31) has argued that land called *Blackdyk* in 1396-7 may refer to a town ditch, licensed in 1260 but rarely mentioned later.

In attempting to determine a moat's course, reference to two aspects of its construction should be made. In excavating the path of the moat large quantities of soil would have to be removed. Typically this material would be used to embank the moat, or alternatively, employed as a foundation for the adjoining wall, thus raising the wall's height. Either action would provide additional evidence to the course of the moat. However, the soils of the Great Yarmouth area tend to be shifting blown sands and the presence of such unconsolidated material would suggest that this would be unsuitable for a wall base: in addition, any moat would rapidly become filled and its route lost, without continuous maintenance. Neither the pictorial town plan of 1588 (see frontispiece and Figure 5.6) nor the Cottonian picture-map of similar age (Figure 5.3) depict a moat. Manship (1619, 3) referred only to the requirement of a ditch in the 1261 charter. Blomefield

(1810, xi, 356), however, went as far as stating 'the building of the wall was succeeded by the sinking of a moat all around the town, over which bridges were thrown at every gate' and 'it was navigable by boats'.

Following an excavation in 1955 (at TG 526 076, see section *3.2.6*), it was proposed that a moat, external to the wall, was 'an integral part of the original defences' (Green, 1970, 114). The method of construction is said to have been dictated by the nature of the sandy substrate (Green, 1970, 116; Carter, 1980b, 303). This substrate is principally of Flandrian wind-blown sands (*Gt Yarmouth, Sheet 162, Quaternary and Pre-Quaternary Geology*, 1: 50,000 Map, 1990). It has been suggested that the moat was floored with a paving 'of large flagstones' and that this continued beneath the wall to provide a suitable foundation for the wall (Green and Hutchinson, 1960, 130 and Fig. 10; Green, 1970, 113-4 and Fig. 2). Below ground level the wall's external facing or revetment of flints set in mortar is thought to have assisted in containing the moat waters. The present author has tried to establish the identity of the flagstone (for its identity could possibly provide a date of erection of the earliest constructed wall). Unfortunately, this has met with no success. It may well be that the occurrence of a 'flagstone' foundation for the wall, which had been 'proved by probing' beneath water, is conjectural. Clarke (1956) had previously reported briefly on the excavation. The reported results of this excavation run counter to two matters raised in the paragraphs above: there is no evidence of removed soil, and Blomefield's belief that the moat was constructed at a time subsequent to the wall.

Based on the evidence of the excavation of limited size just described, the 'flagstone floor' assists in helping to distinguish a moat from any small area of water. Natural water retention in a moat would not have been expected to be difficult for the elevation of the town in the vicinity of the walls currently remains only between two and five metres above sea level. This, in turn, opens the discussion as to the elevation of the area in the past. Green (1970, 114), following Green and Hutchinson (1960; 1965), argued that the land was '12-13 ft. (about 3.75m.) higher' in the 14th century. Arthurton *et al.* (1994), alternatively, state that relative sea level at Great Yarmouth in the 13th century stood between Ordnance Datum and +1m. There are records that the sea reached the town walls in the south of the town, near Blackfriars (as in a storm in 1287), and that attempts were made to reclaim land in this area in 1290 (Page, 1906, 436). In 1302, there was reference to 'the flow and violence of the sea' and its impact upon the precincts of the Franciscan Friars near Middlegate (*Calendar of Patent Rolls*, 1301-1307, 30 Edward I, m. 16d, 86).

Regrettably the number of excavations that might have revealed further evidence for the presence of an early body of water, or a moat, immediately outside the town walls is limited. That of Rye (1972) is described in section **5.4**. McEwen (1982) briefly described a medieval ditch uncovered in 1978 to the north of St Nicholas Road (TG 5249 0793) and in the area of Priory Plain. The outer edge of the ditch occurred about 10m. from the line of the town wall. All finds, however, were more recent than 1580. Wallis (1995, 10) examined the line of a new pipe trench which crossed the line of the town wall in Lancaster Road. Immediately east of a large piece of fallen masonry interpreted as part of the wall, a 'cut feature', identified from the presence of a distinctive silty deposit at a depth of 2m. was regarded as an external ditch to the wall. The lack of artefactual evidence meant that a date could not be assigned to either the construction or infilling of the feature. It should be noted that in the excavation of Green in 1955 (see above) only 17th century and later infill materials were discovered. Further excavations are clearly necessary. Green (1970, 114), himself, stated that there is 'no historical evidence of a medieval moat'. An examination during May and June 1978, between the North West and North Towers, failed to reveal the former presence of a moat (Anon., 1980, 20). It should be noted that neither Manship (1619) nor Swinden (1772) refer to a moat *immediately external* to the 14th century town wall.

The early manuscript evidence suggests that the Great Yarmouth walls may have had as many as ten gates and 18 towers (Swinden, 1772). Of the ten gates, eight faced towards the east and the North Sea, typically regarded as the direction from which the town would face attack. Apart from the north and south gates, five of the east facing gates were it appears sufficiently wide to take vehicular traffic. These were in each instance gates within towers. From north to south, they were: North, St Nicholas ('long since walled up', Swinden, 1772), Pudding, Market (the widest east facing gate), Oxney's, New (which replaced an earlier gate of unknown width prior to the building of the East Mount about 1569), and South Gates. It is possible that each of the three narrow east-facing gates, none of which occurs in a tower, may have been created to give pedestrian access at some time after the wall's initial construction. Certainly, the last of these, Garden Gate, probably was not established until 1636 (Palmer, 1864a, 114).

If, as has been suggested, a moat existed immediately outside the walls (Green, 1970; Carter, 1980b, 303; McEwen, 1982), a suitable bridge would have been necessary to transport the vehicular traffic for each of the seven major points of access to the town. The bridges would presumably have to be lifted and take the form of a drawbridge, either for defensive purposes or if boats of any size were to use the moat. This would be most simply operated from the adjoining tower. None of the towers involved are illustrated as showing, or known to possess, the fixings or detail for a suitable winding mechanism. The towers, however, are known to have been much reconstructed, both during the 16th century (see section **5.2**) and subsequently, when the evidence for such structures may have been removed.

An early moat adjoining the town wall fails to appear on any plan of Great Yarmouth until 1864. The plan first portraying such a feature was prepared for Palmer

CHAPTER 5 OTHER DEFENSIVE SYSTEMS

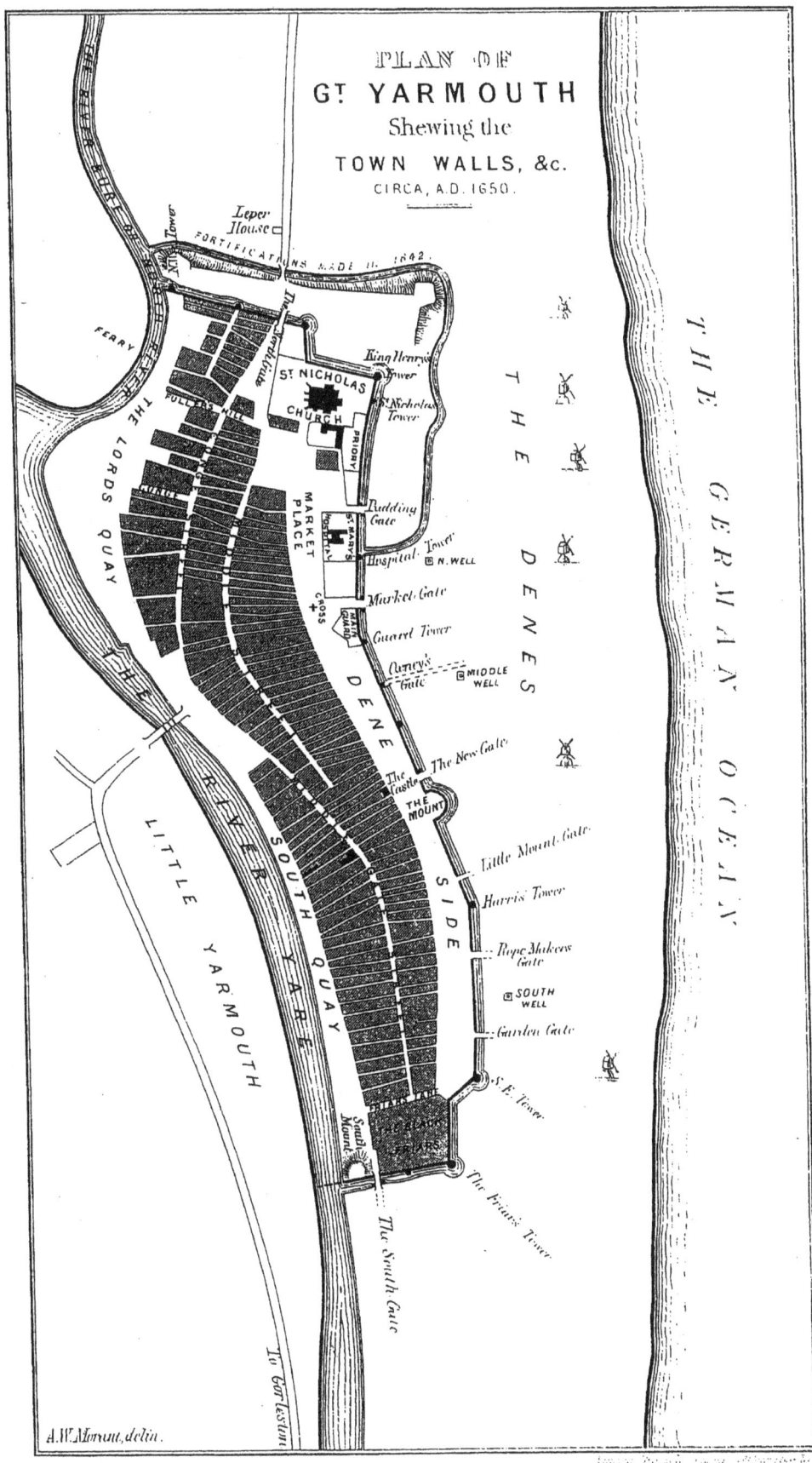

Figure 5.9 This plan was constructed by Mr A. W. Morant for Palmer (1864a) and is supposed to illustrate Great Yarmouth about 1650. This appears to be the only plan which shows a moat adjoining and completely encircling the town walls. It shows the later moat, possibly of 1642, running into it from the north, which on this illustration is from the top of the plan. The plan is interpretive of Palmer's personal views regarding the presence of a moat.

THE MEDIEVAL TOWN WALL OF GREAT YARMOUTH, NORFOLK, U.K.

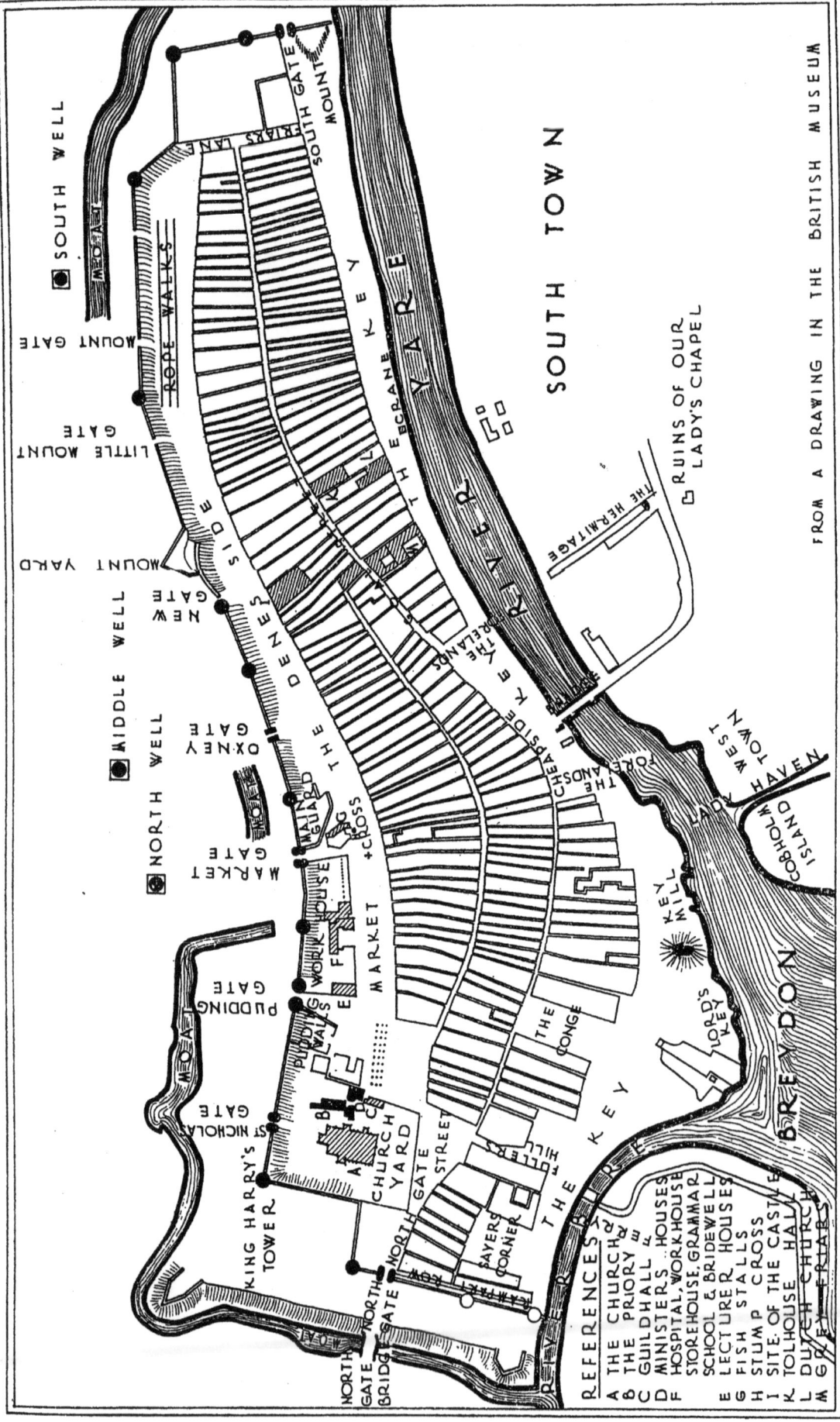

Figure 5.10 A plan originally thought to date from 1619 (Ecclestone and Ecclestone, 1959) and reproduced here from that privately printed volume. It is now thought to have been prepared by R. Cory and to date from about 1820 and be based on 17th century information. The plan clearly displays the moat at the north end of the town (to the left of the plan) and well outside the walls.

(1864a) by A. V. Morant and is supposed to represent the town about 1650 (Figure 5.9). One immediate conclusion regarding such a moat can be reached when it is viewed on a plan: that its many sharp bends in following the wall would make it unsuitable for navigation in anything other than very small boats.

The existence of a 14th century moat or ditch adjacent to the town walls, on the evidence currently available is in the present author's opinion unsubstantiated. The flint facework to the walls has since the wall's inception been subject to frequent repair. Repairs have generally involved the reuse of the recently fallen flints (see section *4.4.2*). Such repairs would have been extremely difficult to undertake had the wall been bordered by a ditch or, more especially from a moat.

The arguments put forward in this section, against the presence of a moat immediately outside and adjoining the town walls, are listed in Table 5.1. To summarise, whereas a ditch possibly containing water in the wet weather, might have adjoined the wall; a moat, particularly one navigable by boats, would appear to have been most unlikely.

5.5.2 A 16th century moat

The case for a moat built about 1588 is substantial. Manship (1619, 74) stated that in 1588, 'a ditch without the south walls (was) digged, delved, and trenched'. Elsewhere, unfortunately, he suggested (p. 68) that this 'mighty main ditch, in former times (was) passable with boats and keels'. The vagueness of this last statement makes it impossible to be aware of whether he was referring to a 14th century moat or one created well outside the walls in 1588. According to Palmer (1864a, 108, supposedly quoting Swinden, 1772; refer also to Rye, 1972), about 1588, the ditch surrounding the wall was 'made passable for boats and keels'. The similarity of these two statements suggests confusion, and Palmer (see 1854, 276), and certainly others, assumed that the moat in question was, at least in part, adjacent to the walls. Had such a moat been accessible to keels and boats with sails, the bridges would have required sufficient elevation for navigation purposes. Swinden (1772, 129) actually wrote ' that there be a ditch or moat made before the north gate walls of the town, sixty feet (18.3m.) wide and eight feet (2.4m.) deep'.

During the Civil War, which broke out in 1642, an *existing* moat was cleaned out (*Great Yarmouth Town Wall*, 1971, 4). Palmer (1854, 277) makes reference to this moat 'surrounding the town walls as far as Pudding Gate', but makes no mention of it following a previous course. Faden's Plan of Great Yarmouth of 1797 (Figure 4.3) clearly illustrates such a moat. This is, however, between 70 and 165m. to the north and north-east of the town wall. It is also illustrated on the plan (Figure 5.9) provided by Morant in Palmer (1864a) and on several later plans. Manship's observation (1619, 74) referred to a moat to the south of the town, and Blomefield (1810, xi, 357) also referred to this 'moat without the south walls'

built in 1588. This moat does not appear on Faden's Plan (Figure 4.3), but a rare indication of its partial presence can be seen on (Figure 5.10) described below (section *5.5.3*). It is probable that the southern branch of the 1588 moat, being less important for defensive reasons, and in a more exposed area of the promontory, rapidly succumbed to infill by blown sands.

After the Restoration in 1660, the requirement for a moat is said to have disappeared and it is described as being rapidly filled with earth and refuse. Carter (1980b, 304) somewhat differently suggested that the 'ditch' on the north side of the town was last cleaned out in 1644 and subsequently 'deliberately filled in'.

5.5.3 Summary

Until such a time as when the existence of stretches of 14th century moat outside and adjoining the town walls are proved by excavation, the view that a moat formed an integral part of the wall structure must remain unproven. With the town and walls closely surrounded on three sides by the sea and rivers the only means of direct attack overland would have been by means of the isthmus from the north and from this direction additional protection would have been advantageous. The statement referring to a new deeper moat in 1642, replacing an existing moat which was cleaned out could readily be interpreted as evidence that any earlier moat or ditch also occupied the same line, and provided defence against attack by means of the isthmus. It seems probable that the earlier moat dated only from 1588, but it would not have been impossible for this to have followed an earlier 14th century cutting.

Earlier written evidence of a moat to the south end of the town is provided by Manship's statement that this was constructed in 1588 (Manship, 1619, 74; Palmer sidelines the statement with the observation 'The South Moat re-trenched'). A drawing held in the British Museum and reproduced by Ecclestone and Ecclestone (1959, 107; Figure 5.10 herein) has been suggested as dating from about 1619. Unfortunately, it appears more likely that the plan was produced in 1820 by R. Cory although it may have been based on 17th century information. This displays a short stretch labelled 'moat' some distance east of Guard Tower separated from a continuous moat, again distant from the walls, extending from about Garden Gate to the River Yare. Possibly in 1820 the outline of an earlier moat was still visible.

The 'moat' is certainly something of an enigma and the evidence from the various town plans suggests that, when present, it was always some tens of metres outside the walls. The shifting nature of the sands of the Denes must have constantly choked its course and made it generally unsuitable for navigation purposes.

Table 5.1 A summary of the points made in the text which may be offered against the presence of a moat immediately beneath and outside the town walls.

1. With the North Sea, and Rivers Yare and Bure closely bordering the town site, only the narrow isthmus to the spit did not afford protection by water.
2. The 1261 licence was to enclose the town with 'a wall and ditch'. The term ditch can also, however, be interpreted as moat.
3. There is no physical trace of either a moat or of the excavated material from a moat in proximity to the wall.
4. There is no indication of a moat (or ditch) beside the walls on either of the two 1550s town plans.
5. The earliest written reference to a moat appears to be in Swinden (1772), but this does not indicate that it is adjoining the wall.
6. Blomefield's record (1810) suggests that the moat was built subsequent to the wall.
7. No trace of the flagstones which Green (1970) assumed to floor both the moat and the wall can be found.
8. Little variation in the land/sea relationship is now thought to have occurred since the 13th century (Arthurton, *et al.*, 1994). Sea surges were known to reach the walls in the past, as in 1287 and 1302.
9. Excavations close to the outside of the wall have so far failed to reveal any artifacts of a date prior to the 17th century.
10. Each town gate would have required a drawbridge to provide a facility for crossing a moat. There is no evidence of any lifting mechanism in any of the preserved illustrations.
11. The first plan to illustrate a moat adjoining the wall is that of Morant in Palmer (1864a).
12. Such a moat would only be navigable by small boats, for the line of the wall possesses too many sharp bends.
13. The flint facework on the external wall surfaces has very frequently been repaired, as far as can be determined reusing the fallen material. If a moat ran below this wall both the repairs and the reuse of the fallen material would have been difficult.

CHAPTER 6

THE FINANCIAL PROVISION FOR WALL CONSTRUCTION

6.1 Introduction

With the onset of the 13th century the move throughout England towards fortifying towns with walls became significant. At times of potential national or international crisis the defence of the towns was thought to be critical both to the king and for the realm. An important coastal port such as Great Yarmouth, therefore, figured relatively early in the new wave of defensive town wall construction. The cost of such an operation was considerable, but the stimulus to build defences in England was sufficient for a new system of providing financial support - the murage grant - to evolve. Turner (1970, 14) defined a grant of murage as 'the permission of the king to levy toll on a specified number of goods coming into the town for sale, the proceeds of which were to be devoted to the construction of the walls'. The terms of the 'grant' (items to be taxed and duration of this local tax), although fixed by the king, were supervised, collected and spent by the town authorities. Murage became the principal source of income for most towns wishing to construct, extend or repair walls. Great Yarmouth, as other towns, also relied on various monetary gifts including, as in 1457 (*Calendar of Patent Rolls,* 1452-1461, 387), relief from fee-farm payments (as custom duties and fines) to the king. Local murage taxes, on first impression, appeared to be an ideal method of covering the costs created in building town walls. They failed, however, to provide a regular income making it difficult to budget for a sustained building programme, they proved complicated to administer particularly in towns where the number of taxable goods became extensive, they impacted on trade, and they particularly taxed those outside of the walled town who wished to bring their wares or produce to market (Turner, 1970). The method of murage payments, therefore, tended to shift by the middle of the 14th century to a tax applied particularly on the more wealthy inhabitants of the town that was involved in fortification. In Great Yarmouth the first instance in which this form of local taxation occurred was possibly in 1369 (Table 6.1).

6.2 Early financial provision – the wall building years

Permission for the town of Great Yarmouth to enclose and fortify the town was granted by a King Henry III charter of 28th September 1261 (*Calendar of Patent Rolls*, 1258-1266, 177; Manship, 1619, 2-3). A grant of murage appears to have been permitted for six years at the same time (Table 6.1) and was made on condition that the townspeople should 'remain faithful to the king and his heirs.' (Swinden, 1772, 78, intimated that a grant of certain customs from merchants trading in the town towards the cost of town walling was made a year before building permission had been granted. This grant, Palmer, 1854, 275, indicated was subsequently withdrawn). Ecclestone and Ecclestone (1959, 84) stated that despite the concession 'nothing happened for about twenty-five years', and Turner (1970, 141) argued 'that a sustained building programme is not likely to have been undertaken until the series of murage grants began in 1321'. Certainly by 1285, when the second murage grant was received, there is evidence of expenditure (*Calendar Close Rolls*, 1279-1288, 328; see sections *1.4.3* and **5.5**). In the intervening period, in 1279 when 'the wall is not yet begun', there had been both a dispute about the levy and a concerned audit of the accounts (*Calendar of Patent Rolls*, 1272-1281, 315).

Records of murage can be traced from two principal sources. There are those records held by central government, frequently documented in the *Calendar of Patent Rolls* but also derived from other central State records, and there are the local town files. The local accounts may provide information on the receipts or the expenditure. From the viewpoint of the present work, details of expenditure are of particular interest for they provide some information on the materials used in the wall construction. Turner (1970, Appendix C) undertook a comprehensive study of all the state documents from which she concluded that Great Yarmouth benefited from 29 (more probably, 27, see below) periods of murage on the following extensive range of dates:

1261-1267, 1267-1272*, 1285-1292;
1322-1329*, 1329-1334*, 1332-1335, 1335-1338, 1338-1341, 1346-1351,1351-1358, 1358-1363, 1363-1370, 1369-1379, 1379-1384, 1384-1389, 1390-1395, 1395-1398, 1399-1403;
1400-1402, 1402-1405, 1406-1409, 1408-1413, 1417-1419, 1419-1423, 1427-1434, 1443-1448, 1457-1464, 1464-1474, 1458-1470.

The three periods asterisked do not tally with Turner's text (p. 139). There is no evidence of the grant listed as occurring from 1267 to 1272 and Turner states that the third grant commenced in 1321 (not 1322 as given in the appendix). No record in the text is given of a grant occurring in the period 1329-1334. Also, some of the 15th century grants do not appear to match the records.

On a number of occasions two murage grants ran in parallel. The dates listed tally closely with those given in Table 6.1 and differences possibly relate to interpretation of the records. In addition to the pavage grant that was converted to wall building in 1399 (Table 6.1), a further pavage grant of four years duration had been given in 1342 (*Calendar of Patent Rolls*, 1340-1343, 425) but there is no evidence that this was used for building purposes.

Ecclestone and Ecclestone (1959, 84) provided information on precisely how murage taxes were applied during the year 1336. As Great Yarmouth was a port of

Table 6.1 Great Yarmouth murage grants 1261-1462. (Compiled largely from English translations of the *Calendar of Patent Rolls*, various dates, HMSO, London: with assistance from Tingey, 1913).

Year	Duration (years)	Reference	Notes
1261	?6	45 Henry III, 28 Sept. m. 3, *Calendar of Patent Rolls* 1258-1266, 177	Tingey (1913, 130). This grant was audited in 1279 (*Calendar of Patent Rolls* 1272-1281, m. 15, 315 when the grant was probably reclaimed; see also Turner, 1970, 139).
1285	7	13 Edward I, 24 June m. 13, *Calendar of Patent Rolls* 1281-1292, 177	Tingey (1913, 133). Granted to take effect 'Michaelmas (= 29 September) next'
?1287	-	?15 Edward I, P. I.	According to Tingey, referred to by Swinden (1772, 78). No record in *Cal. Pat. Rolls* or *Cal. Close Rolls*. Palmer (1856,18) inferred that this might refer to the charter of 1st July 1285 'granted for the purpose of explaining the legal signification of the word "placitet" in the charter of King John'
1321	7	15 Edward II, m. 9, *Calendar of Patent Rolls* 1321-1324, 35	This, according to Tingey (1913, 134) was the third murage grant
1325	7	18 Edward II, m. 2, *Calendar of Patent Rolls* 1324-1327, 134	'In extension of the grant ... of murage for 7 years from 14 November (1321), that, over and above the customs then granted, they shall take for the rest of the term upon every last of herrings entering their port, 2d.'
1327	5	1 Edward III, m. 20, *Calendar of Patent Rolls* 1327-1330, 106	'Further grant to the bailiffs and men ... of murage for five years from the date of the determination of the grant of the same for seven years by letters patent dated 14 November, 14 Edward II (*i.e.* 1320). This should read 14 November 15 Edward II, (*i.e.* 1321).
1332	3	6 Edward III, 19 Sept. m. 10, *Calendar of Patent Rolls* 1330-1334, 333	'Grant ... of murage for three years.'
1335	3	9 Edward III, 6 July m. 32, *Calendar of Patent Rolls* 1334-1338, 151	'Grant ... of murage from 19 September until three years.'
1338	3	12 Edward III, 3 Nov. m. 13, *Calendar of Patent Rolls* 1338-1340, 156	' Grant ... of murage ... for three years.'
1346	5	20 Edward III, 28 Jan. m. 10, *Calendar of Patent Rolls* 1345-1348, 79	Tingey (1913, 134). 'Grant ... of murage for five years.'
1351	7	24 Edward III, 24 Jan. m. 1, *Calendar of Patent Rolls* 1350-1354, 23	Tingey (1913, 134). 'Grant ... of murage for seven years.'
1358	5	32 Edward III, 26 June m. 2, *Calendar of Patent Rolls* 1358-1361, 62	'Grant ... of murage for five years.'
1363	7	37 Edward III, 3 July m. 45, *Calendar of Patent Rolls* 1361-1364, 381	'Grant ... of murage for seven years.'
1369	10	43 Edward III, Part II, m. 6, *Calendar of Patent Rolls* 1367-1370, 334 Granted 24 November 1369	Tingey (1913, 140) indicated without providing evidence that this was to begin July 1370. Reads, 'Grant ... for murage for ten years.'
1369	N.A.	43 Edward III, Part I, m. 4 or *Calendar Patent Rolls* 1367-1370, 256	An order obtained in 1369 (4 June 1369) ... for the repair of the walls, see note * below
1379	5	3 Richard II, 20 July m. 30, *Calendar of Patent Rolls* 1377-1381, 382	'Grant of murage, for five years.'
1384	5	8 Richard II, 12 July m. 38, *Calendar of Patent Rolls* 1381-1385, 438	'Grant ... of murage for five years.'
1390	5	14 Richard II, 28 June m. 39, *Calendar of Patent Rolls* 1388-1392, 277	'Grant ... of murage for five years.'
1395	3	19 Richard II, 30 June m. 28, *Calendar of Patent Rolls* 1391-1396, 603	** 'Grant ... in aid of the enclosing of their town, of certain customs upon things coming to the same for sale, for three years.' (The term murage is not used)
1399	3	22 Richard II, Part III, m. 9, *Calendar of Patent Rolls* 1396-1399, 572	*** Said to be for murage, but *De Pavagio* (pavage money) in margin of roll; 'for thee years.'

Year	Duration (years)	Reference	Notes
1458	10	37 Henry VI, m. 15, *Calendar of Patent Rolls*, 1452-1461, 468	'Grant ... of murage for ten years, to be applied by survey of Walter, bishop of Norwich.'
1462	10	2 Edward IV, m. 3, *Calendar of Patent Rolls*, 1461-1467, 200	'Grant ... of murage for 10 years for the enclosing of the town under the supervision of Walter, bishop of Norwich.'

Notes:
* In 1369; ' to compel all those who have land and rent in the town and all others who stay there continually and live of their merchandise and take profit, to contribute according to their estate and faculties to the repair of the walls of the town ... on pain of arrest'. Tingey (1913, 134) stated that the completed sections of wall were already beginning to show signs of decay, and the ordinary murage grant obtained in 1369 was supplemented by another for the repair of the walls. Tingey (1913, 134-135) cautioned that little heed should be paid to the grave information this message contained, that is, that repair was an imminent necessity, because not merely the town and its neighbourhood, but the whole country was endangered similarly by such lack of wall maintenance (see, Turner, 1970, 83, 87).
** Termed a pontage grant in the margin of the roll, yet 'in aid of enclosing the town' and are silent as regards bridges, and referred to as murage in the calendar of the rolls so assumed to have been used for wall building (Tingey, 1913, 135).
*** Termed a pavage grant in the margin of the roll, yet 'in aid of enclosing the town' and are silent as regards paving, and referred to as murage in the calendar of the rolls so assumed to have been used for wall building (Tingey 1913, 135).
NA Not applicable.

some significance most of the money came from a customs duty on goods entering and leaving the harbour. They stated that 6d was levied on every foreign ship, 1d. in the pound on the value of the cargo, 2d. on each last of imported herrings and 4d on each last exported. A 'last' of herrings was not defined accurately until 1357 and then determined as 10,000 herrings. The authors noted that the highest weekly total was £8 2s. 10d. (at 2006 prices, about £2,400), a week which coincided with the peak of the herring season. These records were obtained from the accounts of the chamberlains which contain both receipts and expenses and are preserved for the years between 1336 and 1345. Turner (1970, 139) indicated that a further set of receipts exist for the three years, 1447-1450 (Borough Court Rolls, Y/C 4/155, 156). Both Swinden (1772, 93) and Tingey (1913, 140 *et seq.*) gave further details of the various goods on which tolls might be levied and the sums exacted for periods in the 13th and 14th centuries when the records are preserved. Table 6.2, which has been abstracted from those records which still remain, reveals something of the typical variation in murage income. Table 6.3, simplified from Tingey (1913, 140-141), provides information on the murage tolls applicable at Great Yarmouth during the period 1327 to 1338. Such details are of secondary interest in that they supply particulars of the imports and exports of the port in question.

Progress on the construction of the Great Yarmouth walls was slow and some portions of the wall were deteriorating well before the entire structure was finished. As Saul (1979, 107) observed, municipal finances were often strained. He suggested that at times of extensive building, as in the mid 14th century, expenditure might be double the murage income. In 1369 (Table 6.1), all who lived off their merchandise and took profit in the town were instructed to contribute to the cost of repairing the defences. As suggested above (section **6.**1) this may have been the first occasion on which the town inhabitants were directly or indirectly obliged to work on the walls, a task described by Swinden (1772, 78) as *murorum operatio* or wall work. Swinden wrote that this work was commuted into money received from the murage taxation or tolls. Ecclestone and Ecclestone (1959, 84) interpreted the procedure slightly differently and stated 'everyone was liable to work for a number of days on the walls. The richer citizens escaped this by paying a sum of money which was used to hire labour in their place'.

Parts of the wall were also in disrepair in 1385 -1386, at a time when there was a renewed threat of a French offensive against England. Everyone owning property in the town was ordered to contribute to the costs and labours of repairing the walls (*Calendar of Patent Rolls*, 1381-1385, 540-541, 545; 1385-1389, 135, 177, 258 and 259).

In the earliest years post the charter of 1209, Great Yarmouth was governed by four bailiffs elected from within the four wards of the town. In 1272, the town was permitted to elect a council of 24 persons to assist the bailiffs. By the early 15th century in addition to these 24 aldermen, the Assembly became supplemented by a further 48 persons who represented 'the whole commonality of the same borough' (Ecclestone and Ecclestone, 1959, 40). This Assembly, actually representing only a small proportion of the town's population, determined the major decisions in the control of the town. The complexities of the murage system, involved the election of officers, muragers, to collect the murage and to determine the patterns of expenditure. In later years, the chamberlains, officers elected to implement the decisions of the corporate body, especially in relation to building and maintenance matters (Higgins, 2005, 5), took over the collecting duty.

Murage income was on occasions supplemented with gifts and legacies provided by the more wealthy citizens

Table 6.2 Typical receipts from murage and monies spent for Great Yarmouth (principally from Turner, 1970, Appendix B).

Year	Grant	Receipts £ s d	Collected over (weeks)	Expenditure £ s d	References
1336–1337	*	–	–	40 9 10½	Swinden (1772, 78-82)
1337–1338	*	–	–	39 2 3	Swinden (1772, 83-87)
1342–1343	–	–	–	72 3 9½	Swinden (1772, 87-90)
1343–1344	–	44 17 10½	25	70 19 3¼	Swinden (1772, 90-91)
1344–1345	–	30 6 1¼	52	88 17 11½	Swinden (1772, 91-92)
1345	–	8 5 0	13	–	Swinden (1772, 93)
1447–1448	*	7 15 11½	–	–	C/A
1448–1449	*	3 10 0	–	–	C/A
1449–1450	–	6 16 7½	–	–	C/A
1457–1458	*	20 0 0	–	20 0 0	E 364/98
1459–1460	*	20 0 0	–	20 0 0	E 364/98

Notes:
* - denotes possession of a grant.
C/A – Great Yarmouth Borough Court Rolls Customs Account.
At July 2006 prices income from murage in 1343-1344 was about £23,000 and in 1459-1460 about £10,500.
Again at 2006 prices expenditure in 1336-1337 was about £20,500 and in 1344-1345 about £45,000.

Table 6.3 Simplified details of those items and goods on which tolls or taxes could be levied under the murage grant authority held by Great Yarmouth over the period 1327 to 1338. Details mainly after Tingey (1913, 140-141).

Item subject to murage levy	Charge
Once a year, on every ship entering the Port of Yarmouth	6d.
On every last of herrings leaving port	2d.
On every last of herrings entering the same port	2d.
On every sack of wool within the port	2d.
On every load (*summa*) of vendible corn leaving port	¼d.
On every garb of steel entering the port	¼d.
On every (*centena*) 50kg. of iron	½d.
On every carrat of lead	2d.
On every 100 pine or deal boards	½d.
On every load of woad	1d.
On every wey of salt	1d.
On every last of ox hides	12d.
On every 1000 of grey work (an inferior kind of fur)	6d.
On every 50kg. of wax	2d.
On every cask of honey	3d.
On every cask of pitch	1d.
On every cask of ashes (an impure form of carbonate of soda)	½d.
On every millstone	1d.
On every 1000 onions	¼d.
On every load of garlic	¼d.
On every trey of seacoals	¼d.
On every 100 oars	2d.
On every bale of bound cloth of value 10 marks or more	4d.
On every bale of bound cloth of value 5 to 10 marks	1d.
On every 100 salted Scotch cod (*mulvell*)	1d.
On all kinds of goods sold by weight of the value of 20s.	½d.
On every cask of wine	2d.
On every boatload of vendible ale within the port	2d.
On every ship laden with ale leaving port, going beyond the sea	12d.
On every 50kg. of copper	½d.
On all vendible merchandise not named, of price 20s. or more	1d.

of the town (Swinden, 1772, 76-78). Blomefield (1810, 354-355) provides such an example when in 1386, the will of a John Rayle bequeathed 'towards the finishing of the walls, 20s. (about £480 in 2006 terms)'.

6.3 Wall rebuilding and modification – the 15th, 16th and 17th centuries

Throughout much of the 15th to 17th centuries the task of maintaining the town walls fell more directly upon the townspeople of Great Yarmouth. In the mid 16th century, for instance, when the Spanish Armada created threats and a state of emergency existed (see also section **5.3**), 'every ward shall cause the whole ward to work three days in the week', this being Monday, Wednesday and Friday, 'with baskets, barrows and shovels to bring in the hills without the walls, to fortify the ramparts for the defence of the walls' (*Great Yarmouth Assembly Book*, 10th January, 1558; as cited by Ecclestone and Ecclestone, 1959, 54). A few years earlier, in 1551, the same threat of invasion imposed the requirement on the inhabitants to no longer 'carry their muck unto the Denes as they were accustomed' but to 'to lay the same on top of the rampart of the walls' (Ecclestone and Ecclestone, 1959, 49). Failing this they would forfeit their barrows. Only butchers were exempt from this requirement. Although she sent munitions and armament, over the same period, Elizabeth I compelled the people of the region to pay considerable sums of money towards the cost of repairing the town walls (p. 86). Manship (1619, 46-47) described his own involvement in the initial erection of the East Mount in 1569-1570 (see sections *3.2.7* and **5.4**) and he elaborated (p. 48) on the costs to the town of the various fortifications, such as the South Mount, over the ensuing years.

From the mid 15th century much more financial support for wall building and repair tended to come from sources other than murage. Fee-farm relief was referred to in 1457 (in section **6.1**). Further similar remissions are recorded, for instance, in 1512 (*Calendar of State Papers*, 1509-1514, 687) and continuing until the 1550s (*Letters and Papers of Henry VIII*, fo. 129). There are records too (see Rutledge and Rickwood, 1970, 7) of elements of remission from general taxes, as in 1541. In 1558, Elizabeth I permitted the town jurisdiction over local maritime laws, which allowed the town's treasury and the walls to benefit from any fines imposed (Ecclestone and Ecclestone, 1959, 39). The same authors also recorded (p. 48) that, in 1551, the ballast of stone, rock or calcyne (chalk rock) of every fishing boat … 'shall be laid ashore … for the building of the town wall at the south end under penalty of 10s.'. These rather drastic measures were cited from the *Great Yarmouth Assembly Book* (Y/C19/1, fo. 20r.; which actually records a penalty of '5s.' and 'bote and fysshez') of that year. Both Swinden (1772, 76-78) and Palmer (1854, 275) stated that money towards the works on the walls, also were received during this period. They again took the form of contributions, bequests and legacies. The more wealthy members of Great Yarmouth's community constituted much of the town's Assembly. On occasions of possible emergency they voted to subscribe money towards the walls and their repair. The *Great Yarmouth Assembly Book* for 21st April 1551, for instance reads, 'it is further agreade that the wall in the south ende shalbe made up as shortely as can be to the buyldyng whereof every four and twentie hath given 4s. 8d. and every of the eight and fortie 3s. 4d.' (Y/C19/1, fo. 15r.). In June of the same year (Y/C19/1. fo. 20r.), the Assembly voted to give £3 of the town's money towards the (re)building of Market Gate. A number of years subsequent to this, in 1570, the walls of the East Mount were in need of repair. Manship (1619, 46-47) records that each of the 'four-and-twenty' gave 2s. every 'eight-and-forty' 12d. 'and the other townsmen according to their ability'.

Occasionally the local records indicate those responsible for the collection of murage and its expenditure and these can be identified by name. For instance, in March 1556, Richard Osborne and Reynolde Turpyn were appointed as the town's muragers. They were appointed to receive the money particularly for 'the making of the town wall at the south end and pay it again as need shall require' (*Great Yarmouth Assembly Book*, 1550-1559, Y/C19/1, fo. 147v., 13th March 1556). In 1551, the town's muragers were Andrew Ives and John Ladd, and in 1653 they were Thomas Muriell and William Harmer. It appears that the major decisions in determining expenditure remained with the Assembly and the preserved Assembly Books reveal much information related to the walls.

6.4 Summary

This brief review provides something of the manner in which moneys were obtained to meet the costs of building, modifying and repairing the Great Yarmouth town wall. The lack of continuous records prevents this information from being clearly associated with the data afforded from the analyses of the wall building materials. The manuscript records do, however, indicate that early construction of the wall, post the 1261 charter, is likely to have been hampered by a lack of continuous financial support as much as commitment. They also reveal the levels of anxiety existing as a result of the potential threats from the Spanish Armada in the mid 16th century and, particularly the concern regarding weaknesses in the south wall, at that time.

CHAPTER 7

RECORDS OF EXPENDITURE ON WALL CONSTRUCTION, MODIFICATION AND REPAIR

7.1 Introduction

Those existing records which provide financial information relating to the Great Yarmouth walls appear to be largely confined to the years between 1336 and 1345, and details contained in the records of the Borough Assemblies from 1550 onwards. These fortunately cover two critical stages in the development of the walls. The first, the early stages in the wall construction and the second the period in which the walls were re-structured and rampired. The accounts for the chamberlains for the early period contain both expenses and receipts. The information can be subdivided into three categories: the provision of bricks; the supply of stones and other materials such as lime for making mortar; and the payments made to construction workers or builders. Table 6.2 shows that expenditure on the walls was extremely variable.

The information provided in this Chapter is largely illustrative and does not represent a thorough trawl of all available documents. Turner (1970, 139) stated that a further set of receipts existed for the three consecutive years 1447 to 1450 but this represents a period when the main building period of the early walls is thought to have been completed. Turner appears to have been referring to the Borough Court rolls (Y/C 4/155,156) which provide murage rates for those years. According to Rutledge (1976, 7) these were discovered behind the wainscot of the guildhall when it was demolished in 1850. A muragers' file containing accounts and inventories of munitions for 1543-1544 also exists (Rutledge, 1976, 26).

7.2 The purchase of bricks

The records compiled from the muragers' accounts of disbursements by Swinden (1772, 79-91) provide information related to the purchase of what are presumed to be new bricks which must have been used in the early years of wall building. These are enumerated in Table 7.1, and cover the period between 1336 and 1345, with gaps. In certain instances the purchases refer to tiles (written as *tegulae*). In the absence of tiles in the modern sense in the wall, these must relate to bricks; *tegulae* being a term duplicating bricks in that period. The 'tiles' and bricks were noticeably of similar price. Yaxley (2003, 120) defines a last as a quantity of 10,000. For herrings, a last was not precisely defined as 10,000 until 1357 (section **6.2**). Originally the term was referred to a load, as a boat or wagon load, and actual quantities or weights varied regionally and according to commodity.

The total expenditure on bricks for the five years listed in the period amounts to £52 10s. 11d., which in 2006 terms amounts to just under £30,000, significantly less than the costs of the present day. In the period, where both number and cost are declared, a total of 410,700 bricks were obtained for £43 7s. 7½d. The cost for a thousand bricks was, therefore, about 2s. 1d. and the number of bricks purchased in 1343-1344 must have been approximately 90,000. These bricks, totalling about half a million in number, would all fall into the Type 2 variety of the present work from their purchase date (see section *2.3.2*). Their distribution in the walls is discussed in section **4.2**. It is just possible that the 1,000 'tiles', 'bought in a certain house' in 1337-1338 may refer to reused bricks of Type 1 category.

Unfortunately, no separate documentation appears to exist relating to the purchases or acquisition of bricks at a later date than 1345. This is particularly regrettable because the majority of the bricks now visible in the wall were manufactured more recently than 1345.

7.3 The purchase of stones and other wall materials other than bricks

The period 1336 to 1345 reveals similar information regarding the purchase of materials like flints as it does for bricks. The information is again compiled by Swinden (1772, 79-91) and is now shown as Table 7.2. Palmer (1854, 275), described the materials in the walls apart from bricks, as former Norfolk flints which were called, 'rock stones mixed,' and 'white rock stones, called calion.' With little or no other early rock material other than flints present in the walls, the 'stones' listed in the Table must all be of flint. The Caen Stone from France was probably incorporated into the wall towers. There was an extremely significant difference between the cost of Caen Stone and flints; a boat load of flints being of similar price to one block of Caen Stone. The Caen material was presumably hewn to shape and possibly ornamented, although from the evidence of the blocks seen in the towers they were not likely to have been large. For the years in which the records are held, the two years 1343 to 1345 were those in which flints were in particular demand. Although none of it may now remain, possibly the earliest wall facing was being added in areas of the wall at this time.

Ecclestone and Ecclestone (1959, 85) recorded that in the year 1336, 13 'treys of lime' cost 4d. The lime was used for mixing into the wall matrix as mortar. This can be compared with information provided two centuries later from the Borough Court Book for 13th October 1551 (Y/C5/4 fo. 19r.) when a John Parfeye received £11 4s. 'for the purchase and carrying of lime' for repairs on the south wall.

CHAPTER 7 RECORDS OF EXPENDITURE ON WALL CONSTRUCTION, MODIFICATION AND REPAIR

Table 7.1 The purchase of bricks or *tegulae* for wall construction purposes at Great Yarmouth over the period 1336 to 1345.

Year	Provision	Total	Cost £	s.	d.
1336–1337[a] (Swinden, 1772, 81)	Bought of Richard Perles 5 lasts* of bricks [50,000], the price of each last £1 3s. and besides for advantage 2,500.	52,500	5	15	0
	Bought of William Webster 3 lasts of bricks [30,000], without an increment, the price of each last £1.	30,000	3	0	0
	Bought of William de Horning 200 bricks, the price 7d.	200			7
	Sum of the bricks, 8 lasts [80,000],1500.	81,500	10	0	8½
1337–1338[b] Swinden (1772, 84)	Bought of William de Horning 2 lasts of bricks [20,000], the price of each last 20s. and 1000 for an increment.	21,000	2	0	0
	Bought by John Neve in a certain house, 1000 tiles, the price 2s.	1,000		2	0
	Bought of William Perles 6000 tiles, the price of each 1000, 2s.	6,000		12	0
1342–1343[c] Swinden (1772, 88)	Bought and received of Richard Perles 6 score 13 thousand and a half of tiles [133,500], the price of each thousand, 2s.	133,500	13	7	0
	Bought and received of William Horning 44 thousand tiles, the price of each thousand, 2s.	44,000	4	8	0
	Also bought and received of William de Childerhouse 1000 tiles, the price 2s.	1,000		2	0
1343–1344[d] Swinden (1772, 90)	The sum of the money disbursed for bricks, £9 3s. 3½d.				
1344–1345[e] Swinden (1772, 91)	Payment for 4 lasts of bricks [40,000]	40,000	4	0	4

Notes:-

a 23 November 10 Edward III [1336]-23 November 11 Edward III [1337]
b 23 November 11 Edward III [1337]-23 November 12 Edward III [1338]
c Saturday after feast of St Peter in chains (1 August) 16 Edward III [1342] - Saturday after feast of St Peter in chains (1 August) 17 Edward III [1343]
d Saturday next after the feast of St Peter in chains (1 August) 17 Edward III [1343] - Saturday in the vigil of St Peter in chains (?31 July) 18 Edward III [1344]
e 1 August 18 Edward III [1344]-29 September 19 Edward III [1345] [14 months]
* According to Yaxley (2003, 120) there were 10,000 bricks in a last.

Table 7.2 The purchase of flints and stones for wall construction purposes at Great Yarmouth over the period 1336 to 1345.

Year (Source)	Provision	£	s.	d.
1336–1337[a] Swinden (1772, 79-80)				
	Bought of Bartholomew de Thorp a certain quantity of stones of his ship's lastage (*lapid' de lastag'* = ballast).		2s.	0d.
	Bought of John Neve two small boats of stones (*lapid'*).		2s.	0d.
	Bought of John de Holm 1 boat of rock stones mixt (*petr' mixt'*).		8s.	0d.
	Bought of John de Holm a boat of white rock stones (*petr' alb'*)		9s.	0d.
	Bought of John de Holm 1 small boat of white rock stones		4s.	0d..
	Bought of a certain stranger, by John Neve, 1 boat of rock stones called calion (*petr' dict' calion*).			6d.
	Bought of Nicholas Hamond one boat of rock stones, called calion.		1s.	9d.
	Also of John Yutte, one boat of rock stones called calion.			7d.
	Bought of William Litster de Acle, one boat of white rock stones.		7s.	6d.
1337–1338[b] Swinden (1772, 84)				
	Bought by the hands of John Neve 2 boats of stones, of which the price of one is 10d. And of the other 2s. 6d.		3s.	4d.
	Two other instances of the purchasing of stones – no detail.			
1342–1343[c] Swinden (1772, 87)				
	Bought ... 5 boats of stones, the price of each boat 12s.	£2	10s.	0d.
	Bought and received of a certain mariner of Cromer a boat of stones, called calion.		5s.	0d.
	Also bought by the hands of William Bennet 5 stones from Caen (*lap' de Cadamo*).		5s.	0d.
	Also bought by the hands of Peter Creffy 7 stones from Caen.		8s.	0d.
1343–1344[d] **Swinden (1772, 90)**				
	The total charge of the stones bought and carried.	£18	14s.	3d.
1344–1345[e] Swinden (1772, 91)				
	Payment for 34 boat loads of stones.	£21	11s	11d

Notes:-

a 23 November 10 Edward III [1336]-23 November 11 Edward III [1337]
b 23 November 11 Edward III [1337]-23 November 12 Edward III [1338]
c Saturday after feast of St Peter in chains (1 August) 16 Edward III [1342]-Saturday after feast of St Peter in chains (1 August) 17 Edward III [1343]
d Saturday next after the feast of St Peter in chains (1 August) 17 Edward III [1343]-Saturday in the vigil of St Peter in chains (?31 July) 18 Edward III [1344]
e 1 August 18 Edward III [1344]-29 September 19 Edward III [1345; a 14-month period]

7.4 Payments made to persons involved in the construction or building of the walls

From the accounts of the chamberlains from 1336, Ecclestone and Ecclestone (1959, 85) cited certain examples of payment to people involved in the construction of the Great Yarmouth walls. Valeniana de Watton, for instance, was paid 3d. for making the treys of lime detailed in section **7.3** into mortar. The masons, William de Weston and William de Setter, were paid £8 5s. 7d. for building 11 rods (about 55m.) of wall, from the foundation to a height of 16ft. (4.87m.) as well as providing 'the cover for a certain gate'.

Swinden (1772, 85-86, 89, 91-92.) also referred to payments made to masons for the erection of stretches of wall to specific lengths and heights over this period. Examples of typical expenditure are given below:

Swinden (1772, 85), in 1337; 'Paid to John Fatherman, John Black, and John Marham, masons, for the building of 13 rods (65m.) 8ft. (2.44m.) high, and the – above. And for 3 rods (15m.) 8ft. high without the – And for 3 rods in the foundation at the gate 11 ft. (3.35m.) high, £10 2s. (about £5,300 in 2006 terms)'.

Swinden (1772, 89), in 1342; 'Paid to William Weston, John de Marham, John Black, and John Almigamen, and their partners, masons, for, the height of the wall, viz. from the middle of the wall to the top, the length of 23 rods and a half (about 118m.), with the rotundity of the tower, and the height of the same tower, £12 (about £6,500)'. The tower in question is thought to be Blackfriars' Tower.

Swinden (1772, 89), in 1342; 'Also paid to the same masons, for cinters made for the same, 1s. 9d. (about £45)'.

Swinden (1772, 91), in 1345 (1st August – 29th September); 'Paid to John Almigamen, mason, for building five rods (25m.) of wall at the bottom, £2 8s. 4d. (about £1,300). And for digging of the foundation of the said five rods, 7s. 6d. (about £185). Also paid to William de Weston and his partners, masons, for the digging of 17 rods (about 85m.) of the wall, £1 5s. 6d. (about £625). And for the building of the said 17 rods at the bottom, £7 10s. 4d. (about £3750)'.

Payments of this nature, to stonemasons would appear to be relatively lucrative when compared with the rates which must have been achieved by persons employed in brick-making, where, during this period, as little as 2s. (£40 to £50) were paid for the receipt of 1,000 bricks (see section **7.2**).

In the 1550s the primary source of information concerning payments is the Borough Assembly records. The necessity for all inhabitants to work for three days a week on the defences during the emergency requirements of 1558 has been described in section **6.3**. Enlarging on an example already referred to in section **6.3**, the Assembly records of the 14th June 1551 (Y/C 19/1 fo. 20r.) may be cited. It reads: It was agreed that 'the towne shall gyve £3 of the townes money towarde the beyldyng of the markett gat and William Mayowe John Echarde Wylliam Heylott and Richard Oldryng in the open house promysed and undertoke to buylde the seid gate able and substancyall with th'elpe of those be of the wards abutt against the same markett ...'. In the Assembly Book records, in certain instances, the task is detailed, but the payments involved are not provided. A typical example occurs on the 28th February 1556 (Y/C 19/1 fo. 146r.) which relates to the 'makying of the sowth wall by the late blacke fryers'.

Manship's personal participation in the building of the East Mount has been referred to in sections *3.2*.7 and **5.4**. He also provided (1619, 47) details of the some of the costs involved in improvements to the Mount in 1588 when labour of the town's inhabitants was 'rated at eight pence a-day' and amounted to more than £200 in total. Similar use of townspeople on the smaller south mount in the same year cost £100 (p. 48).

CHAPTER 8

SUMMARY – RESERVATIONS AND CONCLUSIONS

8.1 Introduction – the project

This work originally commenced as a relatively simplistic study. Great Yarmouth town wall provided a reasonably well-preserved early historic structure in which the range of geological materials was not particularly extensive. Previously unexamined, the geological composition and analysis of these materials was expected to afford both confirmatory and supportive detail to the already recognised and published widespread information and knowledge that had been gleaned from historic manuscripts and documentation.

Initial exploratory examination revealed that although the range of wall fabric material was small, the structural relationships within the wall of these materials were extremely complex. The walls had been rebuilt on numerous occasions. The core of the wall had suffered far less alteration than its facings and especially its external facing: the three components of the wall, internal, core and external surface, required separate analyses. The external facing surfaces of the wall had clearly required frequent repair at different times and were extensively patched. The relative indestructibility of the flints used in this facing material assured their reuse rather than replacement at times of patching or rebuilding. A wide range of different rock types, in this work collectively described as 'exotics', had been recognised initially in certain areas of the external wall face (Table 4.3). These, obtained either as ballast from ships or of glacial and fluvioglacial derivation, were, with study, expected to offer additional information regarding features such as economic trading patterns to the early, as 14th century, history of the town. It was to be discovered that these rock types had been added to the wall faces in recent centuries, any ballast material reflected, therefore, trading patterns only of the relatively recent past.

The core and inner surfaces of the wall were observed at first examination to contain very large numbers of bricks; if as was generally claimed the wall dated to the early or mid 14th century, these would have represented one of the largest collections of bricks of this age in the country. It became apparent that instead, the bricks were of different ages and types, reflecting rebuilding. The larger proportion of the bricks employed in the walls proved to be broken, again indicating probable reuse.

With these aspects of the wall fabric, together with other geological and similar information, understood; it obviously proved necessary to make a fairly far-reaching trawl of the existing published historical records. The extent to which the evidence from the wall fabrics related to the known historic records of the wall required an answer.

8.2 Answers?

The historical records reveal that approval to build town walls was given to Great Yarmouth in 1261, and substantial documentary evidence exists regarding the acquisition of suitable building materials such as bricks and flints in the period 1336 to 1345. Precisely when the walls were completed to their full height, with wall walk, battlements and towers intact is not, however, known. Certainly the quantities of material recorded in the documented period would have been insufficient to complete the full wall. It appears likely that the walls were erected in stages, and as suggested by Turner (1970, 139-141), to various levels in different parts, these probably determined by the material supply and the priorities deemed necessary at the time. As a result, the first repairs to walls are recorded in 1369 (*Calendar of Patent Rolls*, 1367-1370, 87). The latest proposed date for the completion of the walls has been stated as 1396 (Palmer, 1854, 275; section *1.4.4*), but the last recorded murage grant did not terminate until about 1472 (Table 6.1).

Although Green (1970) offered his views on the walls' construction a number of important unanswered questions remain. Various inherent problems exist in constructing a moat as suggested by Green immediately alongside a particularly high defensive wall, especially when the substrate for both is no more than sand and gravel. The civil engineering problems alone are significant. The one flint thick, external wall facing from the evidence of the present study was frequently damaged, or more probably it simply collapsed. The fallen flints were it seems, as might be expected, economically reused. From within a moat such a task, both in gathering the flints and rebuilding would have been difficult.

The regal requirement of the 1261 charter was for a wall and ditch. To the term 'ditch', the interpretation moat may be given. Typically this would have involved excavating the ditch and building the wall upon the excavated material which had been placed on the internal side of the ditch. In Great Yarmouth's case the excavated material, being principally of sand, would have been unsuitable as a wall foundation. Instead the foundations for the town's walls were placed in an excavated trench as is evident today. Although some of the excavated material may have been placed to the interior of the wall much appears to have been carted away from the site of the walls and disposed of elsewhere.

There is apparently no historic record of major work being pursued on the town walls subsequent to 1472 (the date of the last murage), 'until after the 1539 inspection' of the east coast defences was undertaken (Carter, 1980b, 303). Around this date the walls began to be restyled to meet new advanced methods of warfare utilising cannon. Through the mid 16th century until perhaps as late as 1588 (Manship, 1619, 46-48, 73-74; section *1.4*.5) rampires and earth mounts were erected to strengthen the defences. The internal rampires must have exerted considerable lateral thrusts upon the walls and ensuing wall weaknesses must have resulted in the walls requiring some rebuilding. The amount of this rebuilding from the evidence of the present study appears to have been far more extensive than is recorded.

A moat, or stretches of moat, some distance outside of the walls, had been constructed some time prior to the 16th century. At the times of national emergency, in 1588 and 1642 (section **5.5**) this moat was re-excavated. Whether the moat had pre-existed over the full length of the excavation and how long it had previously existed remains unknown. The fragments of moat illustrated on the 17th century maps *might* possibly follow a similar course to the 'ditch' first being constructed in the early 14th century.

Subsequent to the Civil War both the moat and the wall fell into disuse. Carter (1830b, 304) states that the moat was last cleaned out before the termination of this war in 1644. It was subsequently deliberately filled. Cannon remained on the walls until the 1680s, but domestic and industrial town buildings were to subsume much of the wall's surface over the next century.

8.3 Evidence from the wall fabric survey

It must be emphasised, that less than two-thirds of the walls have actually ever been observed since the date of their first construction. If the limited excavation undertaken by Green (1970) is typical of the whole wall circuit, nearly 40 per cent of the wall is below ground. Estimates of material quantities, if undertaken, would have to consider this important hidden element of the structure. Excavations for the wall foundations appear to have been of the order of 3.5m. deep. The unconsolidated sediments of the area would have necessitated extensive timber shuttering if the excavations were not to collapse.

The knowledge attained as the result of the present study of the Great Yarmouth wall materials, on initial brief review, is difficult to relate closely to the historic information regarding the walls. The historic details are restricted only to those periods when documentary information has been discovered. Analysis of the records (see Table 1.1) reveals that subsequent to the permission for the wall being granted in 1261 the historic particulars regarding wall construction or alteration are limited only to restricted periods. With minor exceptions these fall between the years; 1336-1400, 1539-1590 and 1626-1642; in total less than 20 per cent of the wall's entire history. Each period can be associated with a time of potential national crisis: the Hundred Years War with France, the possible invasion by the Spanish Armada and the Civil War. Historically, the details advise that the wall was built in the first period, comprehensively modified in the second to meet new styles of warfare and adjusted to meet the demands of additional ordnance in the third.

In practice, the wall together with the other defences such as towers, gates and mounts (Figures 8.1 and 8.2), must have proved a continual burden on the town's finances. No doubt, from the author's observations over the last few years, the current Borough Authorities would certainly concur with this observation. In particular, the wall's construction, involving a relatively thin (one flint thick) flint facing must have always necessitated continual repair. This is borne out by the fabric survey which revealed the very extensive amount of surface patching (sections *4.4.1* and *4.4.2*). The repair has been such that it is doubtful if any of the originally constructed flint facework remains, except below ground level.

The continuity of repair to the wall is also portrayed by the bricks which are included in the wall. Reused bricks always occur in abundance and these seem to cover in their types an almost continuous span of ages of manufacture, ranging from bricks made prior to the wall being authorised to the present day. However, for most stretches of wall a last date of possible construction can

Figure 8.1 This pen and ink drawing of the North West Tower viewed from the River Bure to the west, was completed about 1955 by Noel Spencer. It is displayed in the Time and Tide Museum from where it was kindly reproduced.

Figure 8.2 The South Gate: an etching undertaken in 1812 by John Sell Cotman. In that year the gate is said to have been demolished. Compare this figure with Figure 3.67 published in 1819. This figure was reproduced by kind permission of the Norwich Castle Museum and Art Gallery.

be offered, this determined by the brick types included (see Tables 4.1 and 4.2, Figure 4.1, and section **4.2**). Two significant, approximate building dates were discovered; 1450 and 1560. The first of these dates was unexpected, in all instances it appeared to be caused by repairs to the wall, for in every case earlier Type 2 bricks (manufactured 1300-1350) were also present in the wall. On occasions, as seen where the wall had recently been broken through to expose the central core, near St Nicholas School, the earlier Type 2 bricks greatly predominated over the later bricks of 1430-1450 date (Type 3). For the greatest part of the wall the internal wall core was not visible. It would seem very probable that unaltered wall, that had not been repaired, would be present in these stretches but had yet to be revealed. Despite this, it must be concluded that parts of the wall first built in the early 14th century had fallen into decay within the first century or so of their history and by mid 15th century were in need of repair. Substantial repairs are known to have been carried out as early as 1369 and again during 1385-1386.

The second major wall construction date of around 1560, coincides with the historically recorded period of major wall alterations associated with the erection of rampires and mounts. The process of rampiring or embanking the inner walls clearly resulted in considerable additional wall failure which necessitated wall strengthening and repair. Areas which illustrate this are shown on Figure 4.1. With the exception of one part of wall stretch 'J' all other exposed wall core localities south of Pinnacle Tower appear to have required rebuilding during this period. Some of this rebuilding is recorded historically. The written records advise of the repairs required to the East Mount in 1569 (Manship, 1619, 46) and the collapse of the wall between Friars Lane and the South East Tower (Figure 8.3), in 1557 (Palmer, 1852, 390). Palmer (1864a, figure facing 110) suggested that the rampire was constructed to a height of at least 11ft. 6in. (3.5m.), sufficient for a wall of limited thickness without constructed weep holes to collapse following any period of rain. The rampiring, of which much still remains, must also contribute to the frequent failures of the wall's flint facework. Damp from the rampire behind the wall must seep through the lime mortar to dislodge the outer flint face.

Aspects of the wall's history have also been gleaned both from the reused stones from earlier buildings and the various 'exotic' rocks. The source of the Caen Stone in

CHAPTER 8 SUMMARY – RESERVATIONS AND CONCLUSIONS

the East Mount, and the Caen and Barnack Stones and *Viviparus* limestone in the South East to Blackfriars' Towers wall stretch can be related to previous buildings and their prior destruction. The stones incorporated into the wall to the north of Blackfriars' Tower being quarried from the ruined Blackfriars' church and monastery, for example, during the 1550s.

It was concluded that the exotic rocks, all of which are on the exterior and interior surfaces of the walls, had been used only in the last few centuries (section *4.3.2*). They assisted in determining which wall surfaces, over much of these more recent centuries, had been enveloped with buildings. They had been derived in two different ways, as ballast from overseas, and as cobbles and boulders collected from relatively local glacial and fluvioglacial deposits. In both cases, much more intensive studies could have been undertaken to determine in detail the exact original source of the rock types involved. It is recognised that detailed studies of ballast materials, for instance, do give some evidence of commercial trading routes.

This would have required sampling the many rock types displayed and subjecting each to microscopic and similar examinations. As the early history of the wall could not be further elaborated by such studies, and sampling would have required authority, these examinations were not undertaken.

Figure 8.3 This watercolour of the rear of the South East Tower was also completed by Cotman in 1812. It is again kindly reproduced from the Norwich Castle Museum and Art Gallery collections. In 1812 the rampire had suffered little destruction.

Figure 8.4 The River Bure, the North West Tower and the old spire of St Nicholas Church, all figure in this delightful scene painted by W. H. Hunt in watercolour, in 1860. This figure was copied, with kind permission, from the original which hangs in the Time and Tide Museum, Great Yarmouth.

In attempting to establish the construction history of the walls, the inner core of the walls and their variety of brick types provide the greatest information. It is important that a watching brief be maintained on areas of future exposure of the inner brick composition. The case of the St Nicholas School wall exposure, created in recent years, emphasises how much valuable information can be revealed. A comparable exposure would hopefully reveal only Type 1 and 2, or just Type 2, bricks to confirm the walls mid 14th century origins. Similarly, in excavations that reveal portions of the wall that initially might have been below ground level, attention should be paid to its composition, its brick types and flintwork styles.

-0-0-0-0-0-

St Nicholas Church, Great Yarmouth's parish church (Figure 8.4), has always figured prominently in the history of the town. Established in 1119, it clearly influenced the line of the town wall (see section **5.1**). The following reference to this large church might, therefore, be included in this work to bring it to a close. It was most kindly brought to the author's attention by Paul Rutledge.

A diarist visiting the town in 1757 saw this verse scratched on an inn window –

> *The Yarmouth Girls are one and all*
> *Strait as their Steple, tho' not quite so tall*
> *No prouder than thir Priest,*
> *Nor sounder than thir Wall.*

To which someone else had scratched the reply –

> *To scandalise the Ladies this was meant,*
> *But, witty Sir, you missed your Intent,*
> *The Steeple's truly strait, the Spire is bent.*

Phillipps (ms 7258), *Norfolk Record Office*

GLOSSARY

A number of words in this work are not used with great regularity in archaeological literature. They are generally described when they first appear in the document. They are defined briefly again here to enable the casual reader to more readily appreciate their meaning.

CINTER, CYNTER, CYNTRE: The timber frame built to support an arch in the early stages of its construction.

DENE, DEN, DOWN: (Probably from the Dutch word *dunes*, a plain; Palmer, 1854, 313): Common land on sand dunes outside the walled town of Great Yarmouth. In the period discussed in this work used for drying fishing nets, the erection of a number of windmills, obtaining fresh well water and the grazing of livestock.

DIORITE (DOLERITE): A coarse grained igneous rock intermediate in composition to granite and gabbro. It weathers relatively easily and in its weathered state it can be referred to as dolerite.

FACE: The largest surface(s) of a typical brick. A depression formed in modern bricks (the frog) in the upper face provides better adhesion for the mortar or cement. See also, stretcher and header.

FEE-FARM: A locality's yearly rent to the crown, gathered in the form of custom duties and fines.

FLINT: A rock that is composed of microcrystalline silica. It can be described as a chert found specifically in the Upper Cretaceous, Chalk Group.

GABBRO: A coarse grained igneous rock, equivalent in composition to the volcanic, finer grained, basalt. These rocks contain more iron, magnesium and calcium, and less silicon, than granite. The rock is, therefore, darker in colour.

GALLETING (GALLETTING, O'Neil, 1953): In Great Yarmouth, the practice of inserting small flakes of flint (gallets) into the mortar between larger flint cobbles in a wall. Elsewhere, as in south-east England, the practice is also known as 'garneting' or 'garreting'. There, other rock types may be used in place of flint.

GNEISS: A coarse grained metamorphic rock (that is, one which has suffered considerable heat and/or pressure). It is typically banded.

HAVEN: The local name applied to a cut made through the Great Yarmouth sand spit by means of which vessels could enter the sheltered harbour of the River Yare. In total there were seven successive havens. The first (Grubb's Haven) was the natural river passage passing between Caister and Great Yarmouth which became silted-up and eventually closed for shipping. This necessitated the construction of cuts or navigation channels through the sand spit to the south of Great Yarmouth town, the first of which was dug about 1347. As each became blocked by the north to south movement of sand along the coast a new haven was constructed. The most recent of these being completed in 1613 (Carter, 1980a, 302) created under the guidance of a Dutchman, Joas or Joyce Johnson, still in use today. Additionally, most authors refer also to the sheltered river harbourage under the name haven.

HEADER: The two smallest faces on the ends of a brick are referred to as headers. The header measurements recorded in the present work are the two shortest dimensions (width and height). See also, stretcher and face.

KEEL: A flat-bottomed boat used particularly for the transport of freight on rivers.

LAST: A load; a weight or quantity varying in amount with the goods and the locality in England. Yaxley (2003) defined a last of bricks or herrings in East Anglia as 10,000.

MURAGE: A medieval toll, levied by royal authority for a certain period, to enable a town to build or repair its town walls. Murage was levied normally on specific goods being brought into the town for sale or merchandise being transported through the town.

NIPPLE(S): A term applied in this work to small markings (of nipple shape) that occur on certain broken flint surfaces. Their possible origin is discussed in the text. (See section *2.3.1*). They are unusually abundant on the broken flint surfaces of the medieval walls of Great Yarmouth.

PUTLOCK (HOLE): A putlock or putlog is a horizontal piece of timber used for supporting wall scaffolding. The putlock was secured at one end in a hole in the wall (at the other, in the scaffolding framework). Traces of putlock holes often remain in walls of any height and they are sometimes reused at the times of wall repair.

RAMPIRE: An earthen ramp built on the inside and against a wall in order to strengthen the wall and protect it against artillery fire. In Great Yarmouth all the town walls were protected in this way in the mid 16th century.

RAVELIN: Strictly a detached strong fortification (French: *ravelin*). Placed in front of town walls it was normally triangular in shape with the rear wall parallel to the town wall to which it could be linked by a narrow raised causeway or a bridge over a ditch (O'Neil and Stephens, 1942, 2).

ROW: In Great Yarmouth, a very narrow lane or alleyway flanked on both sides with houses. In medieval times in Great Yarmouth there were more than 150 such parallel rows all running approximately east-west.

SQUINCH ARCH: A small arch thrown across the angle between two structures. In the context of the present work used between a tower and the adjoining wall walk.

STAPLE: A commodity of prime importance to the economy (such as wool). A town could in medieval times gain royal authority to have controlling influences over the movement (particularly export) of a staple. Great Yarmouth became the staple (town) for wool for a short period and in the later half of the 14th century rivalled Norwich for the status.

STRETCHER: The faces on a brick which include the measurements of the greatest length and the height. See also, face and header.

TREY: A local weight used in East Anglia. Yaxley (2003) describes this as a weight of eight bushels, amounting to about 291 litres or cubic decimetres.

REFERENCES

Relevant material in magazines or articles is cited in the text under the title of the publication (e.g. *Great Yarmouth Town Wall; The Builder*). Note that publications within the Journal *Yarmouth Archaeology* are provided generally only with a year of issue and with no volume number.

Allen, J.R.L. 2007. *Late churches and chapels in Berkshire*. British Archaeological Reports, British Series No. 432, BAR Publishing, Oxford.

Anon. 1980. Rampart Road, Great Yarmouth. *Yarmouth Archaeology*, 1(2), 18-21.

Arthurton, R.S., Booth, S.J., Morigi, A.N., Abbott, M.A.W. and Wood, C.J. 1994. *Geology of the country around Great Yarmouth. Memoir for 1:50 000 geological sheet 162 (England and Wales)*, with an appendix on the Upper Cretaceous biostratigraphy of the Trunch Borehole (Sheet 132) by C. J. Wood and A. A. Morter. Her Majesty's Stationery Office, London.

Ayers, B.S. 1990. Building a fine city: the provision of flint, mortar and freestone in medieval Norwich. In: Parsons, D. (ed.) *Stone: quarrying and building in England AD 43~1525*. 217-227. Chichester.

Ayers, B.S. and Smith, R. 1988. II. The Survey. In: Ayers, B.S., Smith, R. and Tillyard, M. (with a contribution by T.P. Smith). The Cow Tower, Norwich: a detailed survey and partial reinterpretation. *Medieval Archaeology*, 32, 184-207.

Barringer, J. C. (ed.) 1975. Faden's Map of Norfolk. *Norfolk Record Society*, 62.

Blomefield, F. 1805-1810. *An essay towards a topographical history of the county of Norfolk: containing a description of the towns, villages, and hamlets ...*, 2nd ed., (continued by Parkin, C. et al. 11 vols). London.
(Blomefield's account had one of the most prolonged and complex publishing histories of any work produced in the 18th century {Stoker, 1988,17; 1990,124}. The original work was published during the 18th century. Matters concerning Great Yarmouth are typically cited in Voliume 11, published in the second edition in 1810).

Brown, P. (ed.) 1984. *Domesday Book. 33. Norfolk* (2 vols). Chichester.

Carman, W.Y. 1955. *A history of firearms from earliest times to 1914*. London.

Carter, A. 1980a. Great Yarmouth – an introduction. *The Archaeological Journal*, 137, 300-303.

Carter, A. 1980b. Yarmouth – the defences. *The Archaeological Journal*, 137, 303-304.

Clarke, R. 1935. The flint-knapping industry at Brandon. *Antiquity*, 9, 38–56.

Clarke, R.R. 1956. Yarmouth, Great. 78 NE-63/527076. *Norfolk Research Committee Bulletin*, 8, 2.

Clifton-Taylor, A. 1972. *The Pattern of English Building*. London.

Coad, V. 1980. Greyfriars' monastery (TG 525073). *The Archaeological Journal*, 137, 308–309.

Cox, J.C. 1906a [1975 reprint]. The Dominican friars of Yarmouth. In: Page, W. (ed.) *The Victoria History of the Counties of England: Norfolk, II*. University of London Institute of Historical Research, 435-436. *(See note under Cox, 1906c).*

Crowson, A. 1997. Great Yarmouth Town Wall (Site 4294: TG 5255 0671). *Norfolk Archaeology*, 42, 4, 552.

Ecclestone, A. W. 1971. *Henry Manship's Great Yarmouth*. Privately published, Great Yarmouth.

Ecclestone, A.W. 1974. *A Yarmouth Miscellany*. Privately published, Great Yarmouth.

Ecclestone, A. W. 1981. *Yarmouth Haven*. Privately published, Great Yarmouth.

Ecclestone, A.W. and Ecclestone, J.L. 1959. *The rise of Great Yarmouth: the story of a sandbank*. Privately published, Great Yarmouth.

Emery, P. 1998. Evaluation excavation to rear (south) of units 13-15, Market Gates Shopping Centre, Great Yarmouth. *Norfolk Archaeological Unit Report*, 291.

Fakes, A. 2000. Great Yarmouth and the Black Death. *Yarmouth Archaeology*, no volume number, 25-27.

Firman, R. J. and Firman, P. E. 1983. Bricks with sunken margins. *British Brick Society Information Sheet*, 31.

Great Yarmouth Town Wall. 1971. Produced by Great Yarmouth and District Archaeological Society, Great Yarmouth.

Green, C. 1970. Excavations on the town wall, Great Yarmouth, Norfolk, 1955. *Norfolk Archaeology*, 35, 109-117.

Green, C. and Hutchinson, J.N. 1960. Evidence of land and sea level changes from minor excavations in the Yarmouth district. In: Lambert, J.M., Jennings, J.N., Smith, C.T., Green, C. and Hutchinson, J.N., The making of the Broads: a reconsideration of their origin in the light of new evidence. *The Royal Geographical Society Research Series*, 3, 113-146.

Green, C. and Hutchinson, J.N. 1965. Relative land and sea levels at Great Yarmouth, Norfolk. *The Geographical Journal*, 131, 86-90.

Gruenfelder, J.K. 1998. Great Yarmouth, its haven and the crown, 1603-1642. *Norfolk Archaeology*, 43, 143-154.

Harley, L.S. 1951. Essex bricks. *The Essex Naturalist*, 28 (5), 243-254.

Harley, L.S. 1974. A typology of brick: with numerical coding of brick characteristics. *Journal of the British Archaeological Association*, 37, Series 3, 63-87.

Harrod, H. 1855. Notes on the records of the Corporation of Great Yarmouth. *Norfolk Archaeology*, 4, 239–266.

Hart, S. 2000. *Flint architecture of East Anglia*. London.

Harvey, J.H. (ed.) 1969. *William Worcestre Itineries*. Oxford.

Hedges, A.A.C., Boon, M. and Meeres, F. 2001. *Yarmouth is an antient town*. Great Yarmouth Corporation, Great Yarmouth.

Higgins, D. C. 2005. *The ingenious Mr Henry Bell his life his work his legacy his King's Lynn*. King's Lynn.

Hoskins, W.G. 1959. *Local history in England*. London.

Hudson, W. 1907. Norwich and Yarmouth in 1332: their comparative prosperity. *Norfolk Archaeology*, 16, 177–196.

Luard, H.R. (ed.) 1869. Rerum Britannicarum Medii Aevi Scriptores or Chronicles and Memorials of Great Britain and Ireland during the Middle Ages 36. *Annales Monastici*, 4.

McEwen, A. 1982. Priory Plain, Great Yarmouth . *Yarmouth Archaeology*, 4, unpaginated.

McKerrow, R. B. and Wilson, F. P. (eds.) 1958. *The works of Thomas Nashe*. (1904-1910), 5 vols. Oxford. *(Nashe's 'Lenten Stuffe' occurs in vol. 3, 140 et seq.)*

Manship, H. 1619. *The History of Great Yarmouth*. (Printed 1854). The history of Great Yarmouth. Ed. by C.J. Palmer. L. A. Meall, Great Yarmouth and J. R. Smith, London.

(The original copy of Manship's work is held in the British Library and is relatively inaccessible; this copy, printed in 1854,'without deviation or addition' is used throughout the present work for purposes of reference. The page numbers referring to Manship, 1619, detailed in the present work are in every instance those used in the 1854 copy).

Minter, P., Potter, J.F. and Ryan, P. In press, Roman and Medieval bricks: can they be distinguished? *Essex Archaeology and History*.

Nash(e), T. 1599. *Lenten Stuff(e), concerning the desciption and first procreation and increase of the towne of Great Yarmouth, in Norffolke: with a new play never played before, of the praise of the red herring. Fitte of all clearkes of noblemens kitchins to be read: and not unnecessary by all serving men that have short board-wages, to be remembered.*

(Palmer, 1854, 313, describes this book as a 'serio-comic eulogium on the red herring' but complimentary to the town and its inhabitants. See also McKerrow and Wilson, 1958).

Oakley, K.P. 1972. *Man the tool-maker*. British Museum Natural History, London.

O'Neil, B.H. St J. 1953. Some seventeenth-century houses in Great Yarmouth. *Archaeologia*, 95, 141-180.

O'Neil, B.H. St J. and Stephens, W.E. 1942. A plan of the fortifications of Yarmouth in 1588. *Norfolk Archaeology*, 28, 1–6.

Page, W. (ed.) 1906. *The Victoria County History of the Counties of England: Norfolk, II*. Constable, London. *(Pages 435-437 reappeared as three reprints in 1971, under the name of their author: see Cox, J. C. above and below).*

Palmer, C. J. (ed.) 1847. *Greate Yermouthe. A Booke of the Foundacion and Antiquitye of the saide Towne and of Diverse Specialle Matters concerning the same: from the original manuscript written in the time of Queen Elizabeth : with notes and an appendix*. Great Yarmouth.

(Although originally attributed to Henry Manship (senior) by Blomefield, Rutledge (1963) argued that the work was written by Thomas Damet and that he was also probably the originator of the 'Hutch Map').

Palmer, C.J. 1852. Remarks on the Monastery of the Dominican Friars at Great Yarmouth. *Norfolk Archaeology*, 3, 377-393.

Palmer, C. J. (ed.) 1854. *The History of Great Yarmouth, by Henry Manship, Town Clerk, Temp. Queen Elizabeth*. Great Yarmouth.

Palmer, C. J. 1856. *The History of Great Yarmouth, Designed as a Continuation of Manship's History of that Town*. Great Yarmouth.

Palmer, C. J. 1864a. The Town Wall of Great Yarmouth. *Norfolk Archaeology*, 6, 106–124.

Palmer, C. J. 1864b. The Yarmouth hutch, or town chest. *Norfolk Archaeology*, 6, 171–176.

Palmer, C. J. 1872–1875. *The Perlustration of Great Yarmouth, with Gorleston and South-town*. 3 vols., Great Yarmouth.

Pearson, A. 2002. *The Roman Shore Forts: Coastal Defences of Southern Britain*. Stroud.

Pearson, A. and Potter, J.F. 2002. Church building fabrics on Romney Marsh and the marshland fringe: a geological perspective. *Landscape History*, 24, 89-110.

Peterson, J. 2007. Some new aspects of Roman Broadland. *The Bulletin of the Norfolk Archaeological and Historical Research Group*, 16, 23-35.

Pevsner, N.B.L. and Wilson, W. 1997. *The buildings of England. Norfolk I: Norwich and north-east*, 2nd ed., London.

Pictures of Old Yarmouth. 1897. A supplement to the *Yarmouth Independent*, 19th June.

Potter, J.F. 2001. The occurrence of Roman brick and tile in churches of the London Basin. *Britannia*, 32, 119-142.

Potter, J.F. 2004. *Viviparus* limestone ('Purbeck Marble') – a key to financially well-endowed churches in the London Basin. *Church Archaeology*, 5-6, 80-91 and 7-9, 192.

Potter, J.F. 2005a. Field Meeting: Romney Marsh – its churches and geology, 22 May 2004. *Proceedings of the Geologists' Association*, 116 161-175.

Potter, J.F. 2005b. No stone unturned – a re-assessment of Anglo-Saxon long-and-short quoins and associated structures. *The Archaeological Journal*, 162,177-214.

Potter, J.F. 2006. Anglo-Saxon building techniques: quoins of twelve Kentish churches reviewed. *Archaeologia Cantiana*, 126, 185-218.

Preston, J. 1819. *The picture of Yarmouth: being a compendious history and description of all the public establishments within that borough; together with a concise topographical account of ancient and modern Yarmouth, including its fisheries, &c. &c*. Published by the author. Yarmouth.

Rose, C. B. 1860. On the mastoid appearances exhibited on the faced flints employed for the outer walls of buildings. *Proceedings of the Geologists' Association*, 1, 60-63.

Rose, E. J. 1991. The East Mount, Great Yarmouth in the light of recent observations. *Norfolk Archaeology*, 41 (2), 196-202.

Rumble, A. (ed.) 1986. *Domesday Book. 34. Suffolk* (2 vols). Chichester.

Rutledge, E. 1994. Medieval and later ports, trade and fishing, up to 1600. In: Wade-Martins, P. (ed.) *An*

Historical Atlas of Norfolk, 2nd ed., Norfolk Museums Service, Norwich.

Rutledge, E. and Rutledge, P. 1978. King's Lynn and Great Yarmouth, two thirteenth-century surveys, *Norfolk Archaeology*, 37, 92-114.
(See, in particular, Great Yarmouth by Rutledge, P., pp. 110-112).

Rutledge, P. 1963. Thomas Damet and the historiography of Great Yarmouth. *Norfolk Archaeology*, 33, 119-130.

Rutledge, P. 1968. Thomas Damet and his historiography of Great Yarmouth. *Norfolk Archaeology*, 34, 332-334.

Rutledge, P. 1976. *Guide to the Great Yarmouth Borough Records*, Norfolk and Norwich Record Office, Norwich.

Rutledge, P. 1990. Before the walls: the early medieval settlement pattern of Great Yarmouth. *Yarmouth Archaeology*, [no volume number] 41-48.

Rutledge, P. 1991. The will of Oliver Wyth, 1291. *Norfolk Record Society*, 46, 11.

Rutledge, P. 1999. The origins and early development of Great Yarmouth: further evidence. *Yarmouth Archaeology*, [no volume number] 27-33.

Rutledge, P. and Rickwood, D. L. (eds) 1970. Great Yarmouth Assembly Minutes 1538-1545 and the Norwich Accounts for the Customs on Strangers' Goods and Merchandise 1582-1610. *Norfolk Record Society*, 39, 5-80.

Ryan, P. 1996. *Brick in Essex from the Conquest to the Reformation.* Privately printed, Chelmsford.
(ISBN 0 9529039 0 3).

Rye, C.G. 1972. Great Yarmouth – Town Moat. *Norfolk Research Committee Newsletter and Bulletin*, 9, 16.

Rye, C.G. 1973. Great Yarmouth – Blackfriars Church. *Norfolk Archaeology*, 35, 498-502.

Sandred, K.I., Cornford, B., Lindström, B. and Rutledge, P. 1996. *The Place-names of Norfolk. Part Two. The Hundreds of East and West Flegg, Happing and Tunstead.* English Place-name Society, Nottingham.

Saul, A.R. 1975. Great Yarmouth in the 14th century: a study in trade, politics and society. Unpublished University of Oxford D.Phil. thesis.

Saul, A.R. 1979. Great Yarmouth and the Hundred Years War in the fourteenth century, *Bulletin of the Institute of Historical Research*, 52, 105-115.

Saul, A. 1981. The herring industry at Great Yarmouth c.1280-c.1400. *Norfolk Archaeology*, 38, 33-43.

Shepherd, W. 1972. *Flint: its origin, properties and uses.* London.

Stoker, D. 1988. Blomefield's history of Norfolk. *Factotum*, 26, 17-22.

Stoker, D. 1990. Mr Parkin's Magpie, the other Mr Whittingham, and the fate of Great Yarmouth. *The Library*, 12 (2), 121-131.

Swinden, H. 1772. *The history and antiquities of the ancient burgh of Great Yarmouth in the county of Norfolk. Collected from the corporation charters, records, and evidences; and other the most authentic materials.* Printed for the author by John Crouse, 957 pp. Norwich.

The Builder. 1886. vol. 51 (number 2274) (4 September).
(There is no relevant text but there are sketches of two 'Towers on the Old Wall, Great Yarmouth' by Mr G. Ashburner on p. 357.).

Tingey J.C. 1913. The grants of murage to Norwich, Yarmouth, and Lynn, *Norfolk Archaeology*, 18, 129-148.

Trotter, W.R. 1989. Galleting, *Transactions of the Ancient Monuments Society*, 33, 153-168.

Turner, H.L. 1970. Town *defences in England and Wales. An architectural and documentary study AD 900–1500.* London.

Wallis, H. 1995. Great Yarmouth – Caister-on-Sea rising main (Site 30081: TG 524 068 – TG 519 112). Norfolk Archaeological Unit. *Norfolk Archaeology*, 42, 2, 233.

Wentworth, C.K. 1922. A scale of grade and class terms for clastic sediments. *The Journal of Geology* 30, 377-392.

Wilton, J. W. 1979. *Earthworks and fortifications of Norfolk.* Lowestoft.

Yaxley, D. 2003. *A researcher's glossary of words found in historical documents of East Anglia.* Guist Bottom.

ANCILLARY BIBLIOGRAPHY

The texts listed below are not referred to in this document, but they are given here to provide additional related supplementary reading.

Allen, J.R.L. 2004. *Carrstone in Norfolk buildings. Distribution, use, associates and influences.* British Archaeological Reports, British Series No. 371, BAR Publishing, Oxford.

Atkin, M.W. 1983. The chalk tunnels of Norwich. *Norfolk Archaeology*, 38, 313-320.

Ayers, B.S. 1994. *Book of Norwich.* English Heritage, London.

Baggallay, F.T. 1885. The use of flint in buildings especially in the County of Suffolk, *Royal Institute of British Architects Transactions* 1, 105–124.

Bayley, J. 1821. *The history and antiquities of the Tower of London, with memoirs of royal and distinguished persons, deduced from records, state-papers, and manuscripts, and from other original and authentic sources.* Part 1, London.

Benedictow, O.J. 2004. *The Black Death, 1346-1353: the complete history.* Woodbridge.

Betts, I. M. 1996. New thoughts on bricks with sunken margins. *British Brick Society Information Sheet*, 68.

Bown, J. (with contributions by Ayers, B.S., Fryer, V., Huddle, J.M. and Tillyard, M.) 1997. Excavations on the north side of Norwich Cathedral, 1987–88. *Norfolk Archaeology*, 42, 428-452.

Campbell, J. 1965. England, Scotland and the Hundred Years War in the fourteenth century. In: Hale, J.R., Highfield, J.R.L. and Smalley, B. (eds.) *Europe in the late Middle Ages.* 184–216. London.

Clifton-Taylor, A. and Ireson, A.S. 1983. *English stone building.* London.

Cox, J.C. 1906b [1975 reprint]. The Franciscan Friars of Yarmouth. In: Page, W. (ed.) *The Victoria History of the Counties of England: Norfolk, II.* University of London Institute of Historical Research, 436-437.
(See note under Cox, 1906c).

Cox, J.C. 1906c [1975 reprint]. The Carmelite Friars of Yarmouth. In: Page, W. (ed.) *The Victoria History of the Counties of England: Norfolk, II.* University of London Institute of Historical Research, 437.
(Each of these three short articles have been reprinted from Page, 1906).

Cox, J.C. 1914. *The English parish church. An account of the chief building types and of their materials during nine centuries.* London.

Cozens-Hardy, B. 1934. Norfolk Crosses. *Norfolk Archaeology* 25, 297-318.

Cozens-Hardy, B. 1935. Norfolk Crosses. *Norfolk Archaeology*, 25, 319-336.
(Continuation of earlier paper)

Dominy, J. 1995. Bricks and in particular those used at Holkham Hall. In: Longcroft, A. and Joby, R.S. (eds.) *East Anglian studies essays presented to J.C. Barringer on his retirement, August 30 1995,* University of East Anglia, Norwich, 55-57. Norwich.

Dyer, C. 2002. *Making a living in the Middle Ages: the people of Britain 850-1520.* New Haven.

Ellis, H. (ed.) 1859. *Chronica Johannis de Oxenedes.* London.
(A second edition copy, has MS. corrections according to the second edition, and list of errata with the history of the edition, by W.G. Searle).

Fernie, E.C. and Whittingham, A.B. 1972. The early Communar and Pitancer Rolls of Norwich Cathedral Priory with an account of the building of the cloister. *Norfolk Record Society*, 41, 128 pp.

G... L... 1988. A town like Great Yarmouth. Presidential Address [by George Rye] 23 April 1988. *Norwich Research Committee Bulletin* Series 3, 1, 8–10.

Hammond, M. 1991. Bricks with sunken margins. *British Brick Society Information Sheet*, 52.

Hanson, B. 1998. The Dissenters burial ground. *Yarmouth Archaeology*, no volume number, 30-32.

Harris, A.P. 1990. Building stone in Norfolk. In Parsons, D. (ed.) *Stone: quarrying and building in England AD 43~1525.* 207–216. Chichester.

Hart, S. 2003. *The round church towers of England.* Ipswich.

Hatcher, J. 1977. *Plague, population, and the English economy, 1348-1530.* London.

Hedges, A.A.C. 1978. *What to see in Great Yarmouth and district.* Lowestoft.

Hoare, P.G. and Potter, J. F. The medieval masonry walls of Sandwich and their geological composition. Unpublished MS.
(To appear eventually in the comprehensive report of the English Heritage-supported Sandwich Project).

Hoare, P.G., Vinx, R., Stevenson, C.R. and Ehlers, J. 2002. Reused bedrock ballast in King's Lynn's 'Town Wall' and the Norfolk port's medieval trading links. *Medieval Archaeology*, 46, 91-105.

Jondell, E. 1985. Medieval waterfronts in Trondheim. In: Herteig, A.E. (ed.) *Conference on waterfront archaeology in north European towns No. 2 Bergen 1983,* 125–128, Historisk Museum, Bergen.

Kelly, J. 2006. *The Great Mortality. An intimate History of the Black Death,* London.
(First published in 2005 by Fourth Estate).

Kerling, N.J.M. 1954. *Commercial relations of Holland and Zeeland with England from the late 13th century to the close of the Middle Ages.* Leiden.

Knoop, D. and Jones, G.P. 1967. *The Mediæval Mason. An Economic History of English Stone Building in the Later Middle Ages and Early Modern Times.* 3rd ed.. Manchester.

Latham, R.E. 1983. *Revised medieval Latin word-list from British and Irish sources.* British Academy,

London.

Leach, R. 1975. *An investigation into the use of Purbeck marble in medieval England*. Hartlepool.

Lloyd, N. 1925. *A history of English brickwork*. London. *(Reprinted, 1983)*

Mason, H.J. 1978. *Flint, the versatile stone*. Providence Press.

Mee, A. (ed.) 1940. *The King's England: Norfolk: Green pastures and still waters*. London.

Morshead, O. 1951. *Windsor Castle*. London.

Morris, C. (ed.) 1982. *The illustrated journeys of Celia Fiennes c.1685-c.1712*. London.

Nall, J.G. 1866. *Great Yarmouth and Lowestoft*. London.

Naphy, W.G. and Spicer, A. 2000. *The Black Death and the history of plagues 1347-1729*. Stroud.

Parkin, C. 1810. *An essay towards a topographical history of the county of Norfolk*. Vol. 11, 2nd ed., William Miller, London.
(Correct author Blomefield, F, see References)

Pearson, A. 2003. *The construction of the Saxon shore forts*. British Archaeological Reports, British Series No. 349, BAR Publishing, Oxford.

Platt, C. 1976. *The English medieval town*. London.

Reeder, M. G. 1984. Bricks with sunken margins. *British Brick Society Information Sheet*, 32

Rutledge, E. 1988. Immigration and population growth in early fourteenth-century Norwich: evidence from the tithing roll. *Urban History Yearbook 1988*, 15-30.

Rye, C.G. 1962. Midsands Cross, Great Yarmouth. *Norfolk Archaeology*, 33, 114-118.

Rye, C.G. 1964. Yarmouth, Great – TG/525076 Theatre Plain. *Norfolk Research Committee Bulletin*, 15, 12.

Rye, C.G. 1967. Midsands Cross, Gt. Yarmouth. *Norfolk Archaeology*, 34, 240

Rye, G. 1976. Yarmouth outside the North Gate. *Yarmouth Bulletin*. 43.

Rye, G. 1979. Great Yarmouth – Blackfriars Church. *Norfolk Archaeology*, 37, 208.

Saul, A.R. 1984. Local politics and the Good Parliament. In: Pollard, A.J. (ed.) *Property and politics: essays in later medieval English history*, 156-171. Gloucester.

Shrewsbury, J.F.D. 1970. *A history of bubonic plague in the British Isles*. London.

Stoker, D. 2003. Francis Blomefield as a historian of Norfolk. *Norfolk Archaeology*, 44 (2), 181-201.

Stoker, D. 2004. Francis Blomefield as a historian of Norfolk. *Norfolk Archaeology*, 44 (3), 387-405.
(Continuation of article).

Tooke, C.S. 1984. The ballast trade in the port of Great Yarmouth. *Yarmouth Archaeology*, 2 (1), 13-16.

Vinx, R., Hoare, P.G., Ehlers, J. and Stevenson, C.R. 2004. Baltoskandische Glazialgeschiebe in der Stadtmauer von King's Lynn in Südostengland als Beleg einer mittelalterlichen Handelsverbindung. [Baltoscandian glacial erratic boulders in the Town Wall of King's Lynn in southeastern England as proof of a medieval trading connection.]. *Archiv für Geschiebekunde*, 3, 711-720.

www.ingramcontent.com/pod-product-compliance
Lightning Source LLC
Chambersburg PA
CBHW041705290426

44108CB00027B/2857